AF564543

Analytical Techniques and Instrumental Methods in Soil, Plant and Water Analysis

NIPA GENX ELECTRONIC RESOURCES & SOLUTIONS P. LTD.
New Delhi-110 034

Analytical Techniques and Instrumental Methods in Soil, Plant and Water Analysis

Deo Kumar
Assistant Professor
Department of Soil Science & Agricultural Chemistry
College of Agriculture
Banda University of Agriculture & Technology
Banda-210001, Uttar Pradesh

Arbind Kumar Gupta
Assistant Professor
Department of Soil Science & Agricultural Chemistry
College of Agriculture
Banda University of Agriculture & Technology
Banda - 210001, Uttar Pradesh

Atik Ahamad
Subject Matter Specialist (Soil Science)
Krishi Vigyan Kendra, Bharari
Jhansi-284204, Uttar Pradesh

NIPA GENX ELECTRONIC RESOURCES & SOLUTIONS P. LTD.
New Delhi-110 034

NIPA GENX ELECTRONIC
RESOURCES & SOLUTIONS P. LTD.
101,103, Vikas Surya Plaza, CU Block
L.S.C.Market, Pitam Pura, New Delhi-110 034
Ph : +91 11 27341616, 27341717, 27341718
E-mail:newindiapublishingagency@gmail.com
www: www.nipabooks.com
For customer assistance, please contact
Phone: + 91-11-27 34 17 17 Fax: + 91-11- 27 34 16 16
E-Mail: feedbacks@nipabooks.com

ISBN: 978-93-91383-67-1

Composed and Designed by NIPA.

Preface

Efficient management of soil, plant and water is the key to sustained food security and livelihood of mankind. This is particularly true in Indian context, where damage to soil health and soil quality, depletion of soil fertility and degradation of groundwater quality have emerged as major threats to the sustainability of diverse agricultural production systems.

This work contains the details of exercises about analytical determination of physical, physico-chemical and chemical characteristics of soils. In addition, the glassware, stepwise procedures, observations to be recorded and calculation are given for evaluation of results. Some important information about atomic weights of elements, important conversion factors, procedure for preparing standard solutions, optimum range of macro and micro nutrient elements in plants etc; has also been added in the form of appendix. Besides, a chapter in the beginning on laboratory precautions and emergency first aid treatment will be helpful in systematic working in the laboratory and avoiding any accident etc.

A need, therefore, was felt to bring out of this work is to describe the procedures of chemical analysis of soil, water and plant sample in simple way for the use of students, research workers, teachers and persons involved in soil-plant tissue and water analysis laboratories. We have included the procedures for total and available plant nutrients in soil; it will be helpful to understand the changes in nutrient dynamics in different cropping system. We hope that We have succeeded to some extant in this endeavor. Healthy suggestions from learned teachers and scientist for making improvements and enlargement are requested in the next edition will be highly appreciated and honored.

Authors

Contents

Precautions to be Taken While Working in the Laboratory

Working in the laboratory involves the use of chemicals/reagents glassware, items, tools, apparatus/instruments, equipment's and other sophisticated equipment's. To avoid accidents, the following precautions are advised.

Handling of chemicals / reagents:

There are various points to be considered during the use of various chemicals in the laboratory, which are as under-

1. Use apron or laboratory coat.
2. Read the label before opening the container / bottle
3. Check that the chemical is the one, required.
4. All acids are highly corrosive, alkalies and highly caustic. Therefore, use them carefully and do not spill over your clothes, body or on the tables.
5. Open the container carefully in well ventilated area.
6. Seal container tightly after use.
7. Avoid using contaminated apparatus and instruments.
8. Never add water to the concentrated acid, always add acid to water to make required solution.
9. Many chemicals are highly poisonous. Therefore, do not suck them mouth. Use suction bulbs.
10. Do not eat, drink or smoke while handling and using chemicals.
11. Instruments are very costly and their repair is very time consuming. Therefore, handle them carefully. Get the guidance from your instructor.
12. Perform the exercise at seats allotted to you. Do not form groups. Avoid unnecessary any loud talk.
13. Store solutions in air tight glass bottles in order to avoid exposure to air.
14. Store oxidizing chemicals like iodine and silver nitrate only in amber colour glass bottles.

15. Approach your instructor for any clarification and help.
16. Take help laboratory attendant for your requirement of chemicals, solutions and glassware's etc.
17. In spite of these precautions if any mishap occurs during the conduct of the practical, adopt the following safety measures of first aid.

The following precautions will increase accuracy of analytical work

a) Use distilled water for routine analysis and double distilled water for micronutrients analysis.

b) Always carry out blank determination along with test sample.

c) Ensure proper rinsing of pipettes before sucking next sample salutation.

d) Always keep similar conditions and use same reagents both are standard curve and sample analysis.

e) Keep concentration with specific sensitivity range of particular element through suitable dilution in spectrophotometer and atomic absorption analysis.

f) Take adequate precautions to check ingress of moisture during sample preparation and weighing.

First aid emergency treatment in the laboratory

A chemistry laboratory is equipped with different types of chemicals and apparatus. Accident can be occur by chance or due to lack of attention on the part of student. In any case prompt action should be taken to give first-aid to the victim and should be hospitalized, if need be. The probable accidents and their first-aid emergency treatment are given below:

Type of accident	First-aid emergency
1. Burns: i. Burn by dry heat (i.e., flame, hot object etc). ii. Burns causing blisters. iii. Acid burns. iv. Boromine burns.	a. Apply burnol or sarson oil. b. Apply burnol at once. c. Wash liberally with 2% NH_3 solution and then rub glycerin, wipe off glycerin after some time and use burnol. d. Wash freely with water, then with 1% acetic acid and again with water. Dry the skin and use burnol.
2. Cuts: i. Minor cuts ii. Serious cuts	a. Allow to bleed for a few seconds, remove the glass piece if any. Apply a little methylated spirit and cover with a piece of cotton. Alternatively apply Fe CL_3 solution to stop bleeding. b. Apply pressure above the cut to stop bleeding Call the doctor.
3. Eye accidents: i. Acid in the eye ii. Alkali in the eye	a. Wash thoroughly with water, then with 1% Na_2CO_3 solution followed solution followed by water again. b. Wash thoroughly with water then with 1% boric acid solution.
4. Poisons: i. Poisons not swallowed ii. Acid swallowed iii. Caustic alkalis swallowed iv. Inhalation of gases like Cl_2, SO_2, Br_2 etc.	a. Drink a lot of water. Drink lime water. No emetic should be taken. b. Loosen the clothes at the neck. Go in the causing suffocation, open air. Inhale dilute vapors of ammonia of gargle with sodium bicarbonate solution. c. Drink a lot of water. Drink a glass of lemon or orange juice. No emetic should be taken. d. Spit out immediately. Wash the mouth with water several times.
5. Fire: i. Clothes catch fire. ii. Beaker containing inflammable liquid catches fire	a. Do not run. Wrap with a blanket. Lie down on the floor and roll. b. Cover the beaker with duster or damp cloth.

At the same time, one should avoid the following

a. Do not suck dangerous chemicals through mouth. Use section pump.

b. Do not pour large volume of washing solution at the time of washing precipitate but use a number of small portions of washing solution and allow complete draining before addition of fresh portion.

c. Do not heat glass wares directly over flame without wire gauge.

d. Do not heat inflammable chemicals directly.

e. Do not transfer readily soluble substance from weighing bottle in the volumetric flask directly, but transfer it to a beaker, dissolve and then quantitatively transfer in to volumetric flask.

f. Do not keep AR grade chemicals open for longer time.

g. Do not weigh hot or cool precipitate directly as it may absorb moisture but cool it in a desiccator for at least half an hour before weighing.

h. Do not hold stopper between fingers while pouring liquid from bottle nor put in on shelf or working bench. Put in on a clean watch glass.

i. Do not use glass or silica crucible for fusion of alkali metals of evaporation with HF but use platinum crucible (melting point 1773 ^{0}C).

j. Never add water to the concentrated acid, always add acid to water to make required solution.

k. Never open ammonia solution at room temperature. It should be opened after proper cooling to avoid its explosion.

1

Soil Analysis

1.1. Collection, Preparation and Storage of Soil Sample

Soil sampling is the first step of soil analysis. As very small fraction of the huge soil mass is used for analysis, it becomes extremely important to get a true representative soil sample of the field. For collecting a representative soil sample, due consideration must be given on the following:

1. The sample must truly represent the field it belongs to.
2. A field should be treated as single sampling unit if it is appreciably uniform. Generally, an area not exceeding 04 ha is taken as one sampling unit.
3. Variation is slope, color, texture, crop growth and management practices are the important factors that should be taken into account for sampling. Separate samples are required from areas differing in these characteristics.
4. Sampling from recently fertilized plots, bunds, channels, marshy tracts and area under trees, wells, pit, manure dumping sites or other non-representative locations must be carefully avoided.
5. Larger area must be divided into appropriate number of smaller homogeneous unit for better representation.

Sampling Tools and Materials

Samples can be drawn with the help of following the various tools or apparatus-

1. Khurpi
2. Spade
3. Auger
4. Bucket
5. Vernier Callipers

6. Scale
7. Rack
8. Wooden roller
9. Mortar and pestle
10. Sieve
11. Polythene/paper/cloth bags
12. Labels
13. Cardboard cartons
14. Aluminium boxes.

Soil Sampling Procedure

1. Depending on field conditions and the objective of sampling, select proper sampling tools.
2. Based on difference in soil type, colour, crop growth or slope, divide the area in different homogeneous units.
3. At the sampling site, remove the surface litter with *khurpi* or *phawda.* With the help of the sampling tool collect a sample in a bucket. If a *khurpi* is used, make a 'V' shape cut up to 15 cm (vertical) depth and collect the soil. This sample is known as 'primary' sample.
4. Collect such primary samples from different areas of the field in a *zig – zag* manner. Such primary samples should have approximately the same weight.
5. After collecting 20-25 soil samples per hectare, mix the all soil samples in the bucket thoroughly, then crushed or grinds the samples and draws 500 g. to 1.0 kg soil in a clean cloth, paper or polythene bag. This is known as composite sample or ideal soil sample.
6. On two labels write the following information:
   i. Name of the farmer
   ii. Location of the field
   iii. Field number
   iv. Soil depth of which sample is collected
   v. Name of the person who collected the sample

vi. Cropping pattern

vii. Irrigation facility, etc

viii. GPS Location (latitude, longitude and altitude)

7. Place one label in the bag and tie another outside the bag.

Processing of Soil Samples

1. After collecting the sample from the field, spread it on a dry tray or floor. Be sure that the tray has the proper label.
2. Do not keep tray where fertilizer or manure is stored.
3. Do not keep the collected samples in open area or in the direct sunlight.
4. Always keep the sample to dry in shade (on a drying rack) for two to three days, so that the samples become air dry.
5. Using a mortar and pestle or a wooden roller grind the sample and pass it through a 2.0 mm size of the sieves.

Storage

a. Transfer the sieved material in a polythene bag.

b. Do not use bags or boxes previously used for storing fertilizers, salts or its other related chemicals.

c. Put one label indicating details of soil sample inside the bag and finally place it in a cardboard carton.

d. Label the carton properly with the details of soil sample.

Precautions in collection and storage of samples

Special care in collection and handling the soil samples is required for preventing contamination.

Following precautions should be taken to minimize error

1. Avoid contact of the sample with chemicals, fertilizers or manures.
2. Use stainless steel augers instead of rusted iron *khurpi/ kudal* or *kassi* for sampling for micronutrient analysis.
3. Do not use bags or boxes previously used for storing fertilizers, salt or any chemicals.

4. Store soil samples in clean, preferably new, cloth or polythene bags.
5. Use glass, porcelain or polythene jar for long duration storage.

1.2. Determination of Soil pH

Soil reaction (pH) is an important physico-chemical property and it indicates whether a soil is acidic, neutral or alkaline. It is measured and expressed as pH. S.P.L. Sorenson (1909) suggested the term pH (puissance dehydrogen), which means the power of hydrogen. pH is the negative logarithm of hydrogen-ion activity.

Principle pH may be defined as the logarithm of the reciprocal of active hydrogen ions or symbolically:

Whereas, AH is the hydrogen ion activity in moles per litre, A solution with an H activity of 0.001 M will have a pH of 3.0 with an H activity of 0.0001 M, 4.0 and so on. Such measurements are intensity measurements and differ from capacity measurements (such as exchangeable H^+), where absolute amounts are involved. The pH of the soil-water system is related to the ionic saturation of the soil colloidal complex with reference to the ratio of H^+, $Ca^{2+}$and Mg^{2+}. The value of soil pH is affected by several factors, such as:

(i) Soil : water ratio employed,

(ii) Partial pressure of CO_2

(iii) The salt content of soil sample,

(iv) Drying of the soil sample and

(v) The amount of grinding given to the soil.

Soil pH is mostly determined on soil: water (1:2) suspension. The amount of water added varies from enough to bring the soil to saturation to ten times the weight of soil. Schofield and Taylor (1955) have shown that a much reliable measurement of soil pH can be made in a 0.01 M $CaCI_2$ solution. In several European countries, however, soil pH is determined in 1.0M KCl solution.

Apparatus

- pH meter with glass-calomel combination electrode.
- Reciprocating shaking machine.

Reagents

- Potassium Chloride solution, 1 M or Calcium chloride solution of 0.01 M.: Dissolve 74.55 g KCl in distilled water and make up volume to

1000 ml or dissolve 2.18 g $CaCl_2 . 6H_2O$ in distilled water and make up volume to 100 ml.

- Buffer solution, pH 4.0, pH 7.0 and pH 9.2: Dissolve standard analytical concentrate ampoules according to instruction.

Procedure

1. Weigh 25 g soil into 100 ml polyethylene wide-mouth type bottle.
2. Add 50 ml solution (1 MKCl or 0.01 M $CaCl_2$) and cap the bottle.
3. Shake it for 25-30 minutes.
4. Before opening the bottle for measurement, shake by hand.
5. Immerse electrode in upper part of the suspension.
6. Read pH when reading has stabilized (accuracy 0.1 unit).

Precautions in the use of pH meter

1. The electrodes should be allowed to remain in the test solution or suspension longer than necessary.
2. The electrode should always be kept suspended in distilled water.
3. Immediately after use, the electrode should be washed with a gentle stream of distilled water.
4. The pH meter should be calibrated with the buffer in the expected pH range of the soil (commonly pH 4.0 or 9.2 buffers). Buffer adjustments should be rechecked periodically.
5. Never allow sample solution to dry on the electrode.
6. Keep the pH buffer solutions properly stored.
7. Do not allow the electrode to touch the sides or bottom of the container.
8. Read carefully the instrument manual of the pH meter.
9. Do not store the buffer solution for long time.

1.3. Determination of Electrical Conductivity (EC)

Soils contain varying amounts of different soluble salts derived from parent rocks, minerals, fertilizers, irrigation water and decomposition organic residues. When a soil contains an excess amount of salts, it is termed as saline soil. Sometimes it is called “white alkali” soil because of white saline crust that appears on drying. Excessive accumulation of these soluble salts in the soil restrict the plant growth by decreasing the entry of water into the plant roots and hence absorption of nutrients. In arid and semiarid regions, the excessive salts may accumulate due to poor salt water balance under impeded drainage and high water table conditions. It is, therefore, essential that the amount of these soluble salts is determined to assess the salinity problem of soils to decide as to which crops can be successfully grown therein as crops differ considerably to salt tolerance. The tolerance of some important crops to salts in soils is given below:

High salt-tolerant crops	Medium salt-tolerant crops	Low salt-tolerant crops
Barley	Ray	Field beans
Wheat	Radish	
Sugarbeet	Oats	Greenbeans
Rape	Rice	Pea
Cotton	Sorghum	Apple
Kale	Corn	Orange
Spinach	Flax	Grape fruit
Date Palm	Sunflower	
Castorbeans		
Tomato		
Cabbage		
Cauliflower		
Pomegranate		
Grape		

Soluble salts in soils are determined by measuring the electrical conductivity of soil solution. Higher the salt content, more is the flow of electric current through the solution. Electrical conductivity (EC) or specific conductivity is defined as the resistance of a conductor 1 cm long and 1cm^2 in cross sectional area and is expressed as mhos per cm (EC x 10^3) or deci Siemens/m (dSm^{-1}). The soil salinity in relation to plant growth is generally measured in terms of conductivity of the saturated extract of the soils.

Principle

The electrical conductivity of a soil solution is measured with a conductivity meter known as "solu bridge" Ions, like metals, allow the electric current to pass through them. Hence, the electrical conductivity (EC) of the soil-water system rises with increasing content of soluble salts in the soil. Thus, the measurement of EC will give the concentration of soluble salts in the soil at. particular temperature.

Apparatus

1. Weighing balance
2. 100- ml beaker
3. Measuring cylinder
4. Glass rod
5. Conductivity flow cell with automatic temperature compensation.
6. Vacuum line or suction pump.
7. Conductivity meter

Reagent Standard potassium chloride (KC1) solutions, 0.01 and 0.1 N: For 0.01 N solution (1.412 dS/m at 25°C) dissolve 0.7456 g of KCl in distilled water, and add water to make 1000 mL 25°C. For 0.1 N solution (12.900 dSm^{-1} at 25°C), use 7.456 g of KCl.

Procedure

1. Weigh 25 g of soil in a 100 ml beaker and add 50 ml of distilled water.
2. Carry out intermittent stirring with a glass rod for about 30 minutes and then leave it overnight to obtain a clear supernatant solution.
3. Rinse and fill the conductivity cell with standard KC1 solution.
4. Adjust the conductivity meter to read the standard conductivity.
5. Put the "SET/CAL/Read" switch to "CAL/READ" position.
6. Select 20 or 200 dSm^{-1} (mmhos/cm) range.
7. Rinse and fill the cell with soil extract/ soil suspension or water sample and read the C, corrected to 25°C, directly from digital display.

Note: The same soil suspension prepared for determination of pH is also used for salinity determination. After recording the soil pH, allow the soil suspension in the beaker to settle for maximum of 30 minutes (Soil: Water 10 g:20 ml).

Precautions

1. Allow the salt bridge to warm up for 30 minutes.
2. Set the temperature knob to room temperature.

3. Since dilution affects EC measurement, the 1:2 soil: water ratio must be maintained.
4. Allow the suspension to stand for sufficient time to obtain clear supernatant solution.
5. Be sure that electrodes of conductivity cell are completely immersed in the supernatant solution.
6. Wash the cell with distilled water and wipe it dry with tissue paper after each sample.
7. Keep the cell immersed in distilled water when not in use.

1.4. Determination of Soil Bulk Density

The bulk density varies indirectly with total pore space present in the soil and gives a good estimate of porosity of soil. Bulk density is of great importance than particle density in understanding the physical condition of soil.

Soil bulk density is defined as the ratio of the mass of the oven dry soil to its bulk volume.

1. Weighing bottle method

Principle

The mass of the soil is determined by weighing the oven dry soil sample. The soil in small amounts, say 5-6 gm., is placed in a container which is tapped 15-20 times on a table be letting it fall from a height of about 2-3 cm. This tapping is assumed to produce the same packing as occurring naturally in the field, even though this assumption is not strictly correct. The volume of this packed soil will be equal to the volume of the container. Bulk density is calculated from the mass and volume of the soil.

Apparatus

A weighing bottle (50 ml), balance and a burette of 50 ml capacity.

Procedure

Weigh an empty 50 ml bottle. Fill the bottle with oven dry soil upto the brim by tapping and weigh it. Empty the bottle and determine its exact volume using burette.

Observations & Calculations

Mass of empty bottle = M1 gm.

Mass of bottle + soil = M2 gm.

Mass of the soil = (M2 – M1) gm.

Volume of water filling the bottle = V cm3↑

Bulk density = (M2 – M1) / V gm.cm-3↑ or Mg. m-3↑

2. Clod Method

Principle

A few pieces of soil clods are oven dried, weighed and coated with a water repellent substance (rubber solution) just enough to check the water entry into them. A number of such clod pieces are placed inside a graduated cylinder filled with water and the volume of water displaced by them is noted. Knowing the dry weight of the soil and its volume, the bulk density is calculated. The clod must be sufficiently stable to cohere during coating, weighing and handling.

Apparatus and Reagents

Crepe rubber or smoke sheets, benzene, toluene, electric stirrer, wide mouthed container, a thin wire mesh, 250 ml. Graduated cylinder, weighing balance and oven.

Procedure

Method of preparation of rubber solution

Weigh 50 gm rubber sheets. Cut them into smaller pieces and soak in toluene (rubber to toluene ratio is kept at 1:5 by weight). Keep the mixture overnight in a tight container. Next day add benzene (440- 500 ml) and stir the contents thoroughly with the help of electric stirrer so that swollen rubber pieces are shattered and homogeneous solution is obtained. Dilute this solution with benzene to achieve concentration of 1:30 by weight. Instead of benzene, toluene can also be used for dilution but the benzene has the advantage that it dries rapidly when applied over the soil clod. Instead of rubber solution we can also use paraffin wax.

Coating Procedure

Take a soil clod, oven dry it and note its weight. Wrap the clod in a thin wire mesh, dip in the rubber solution placed in wide mouthed container. Remove it momentarily. Repeat the process 3-4 times.

Volume Measurement

Take a 250 ml graduated cylinder containing 150 ml of water. Put the coated soil clod into the cylinder and note the volume of water displaced by the clod. The volume of the displaced water will be equal to the volume of the clod.

Observations & Calculations

Mass of oven dry clod = M gm.

Volume of water in the cylinder = V1 cm3

Volume of water + Clod in the cylinder = V2 cm3

Volume of the clod = (V2 – V1) cm3

Bulk density of the clod = M / (V2 – V1) gm.cm3 or Mg.m-3

3. Core Method

This is a field method for bulk density determination.

Principle

In this method a cylindrical metal sampler or core of known volume is driven into the ground to the desired depth and carefully removed to preserve a known volume of sample as it existed in situ. This core sample is dried at 105°C and weighed. Bulk density is the oven dried mass divided by volume of the sample. The core method is usually unsatisfactory if gravels are present in the soil.

Apparatus

A core sampler, sharp knife, a tray, moisture boxes and oven.

Procedure

Drive the sample into either a vertical or horizontal soil surface far enough to fill the sample but not to compress the soil in the confined space. Carefully remove the sampler and its contents. Trim the soil extending beyond the sampler with a sharp knife. The soil sample volume is the same as the volume of the sampler or the core. Transfer the wet soil to a tray and weight it. Take a portion of the sample in a moisture box, weigh and place it in an oven at 105°C for about 24 hours and weigh it again.

Observation & Calculations

Mass of wet bulk soil sample = M1 gm.

Mass of the moisture box = M2 gm.

Mass of moisture box + wet soil = M3 gm.

Mass of moisture box + oven dry soil = M4 gm.

Mass of wet soil = (M3 – M2) gm.

Mass of oven dry soil = (M4 – M2) gm.

Oven dry mass of bulk soil sample

= (M4 – M2) M1 / (M3 – M2) gm or

say = M5 gm.

Volume of bulk sample / core sampler = V cm3 or $\pi r^2 h$ cm3

Where r is the radius in cm. and

h is the height of the core in cm.

Bulk density = M5 / V gm.cm3

Precautions

Core samples should not be taken in too wet or too dry condition of the soil. In wet soils, the friction along the sides of the sampler and vibrations due to hammering are likely to result in viscous flow of the soil and thus, compression of the sample. When this occurs, the sample obtained is unrepresentative, being more dense than the soil in situ. Compression may occur even in dry soils if they are very loose. Also, hammering the sampler into dry soil often shatters the sample and makes it loose.

Ratings

Textural Class Bulk Density Pore Space (%)

Textural Class	Bulk Density	Pore Space (%)
Sandy Soil	1.6	40
Loam	1.4	47
Silt Loam	1.3	50
Clay	1.1	58

1.5. Determination of Particle Density (PD)

Procedure

1. Add water to an empty graduated cylinder to the 70 ml mark.
2. Transfer 40.0 g of the fine-textured soil into the graduated cylinder.

3. Remove trapped air by carefully tapping the cylinder on the lab bench.
4. Determine the volume of the soil-water mixture.
5. Calculate the particle density. Remember that the volume of interest for particle density calculations is the volume of the soil solids!
6. Use the following formula to calculate the percent pore space.

   %PS = (1Db/Dp)100

   Where: %PS = percent pore space

   Db = bulk density

   Dp = particle density
7. Pour the contents of the cylinder into the buckets in the sink and rinse out any remaining soil material.
8. Repeat steps 1 - 7 with the sandy soil.

$$\text{Pore space\%} = \frac{\text{Particle density} - \text{Bulk density X100}}{\text{Particle Density}}$$

1.6. Determination of Soil Texture by International Pipette Method

Principal

The main principal of this method is to determine the concentration of soil particles at a given depth as a function of time. The concentration of soil particles at depth 'h' and time 't' will be the concentration of particles of different diameter having velocity less than h/t. The different size particles will fall through the fluid at different rates. The time's' needed for important soil separates to fall through a height 'h' is obtained from the Stoke's equation as state below:

$$V = 2/9gr2\frac{(d1-d2)}{\dot{\eta}}$$

where 't' is the time in seconds for a particle to fall in h cm, η is the viscosity of the fluid in poise, r is the diameter of the particles in cm, g is acceleration due to gravity in cm/sec2, d1 is the density of soil particles in gm/cm3 and d2 is the density of fluids in gm/cm^3. Since the viscosity and density of fluid are influenced by temperature, the variations in temperature will cause the changes in rate of fall. So the value of viscosity and density at the prevailing temperature is considered in the calculation.

Apparatus Required

Sieves (2.0 and 0.2 mm), 500 mL tall shaped beaker, water bath, rubber tipped glass rod, shaker or stirrer, 1000 mL measuring cylinder with stopper (graduated length is 36 ± 2 cm), filter paper (Whatman no. 44), Buckner funnel, vacuum pump, special plunger for mixing, stop watch, glass marking pen, specially made mechanical analysis pipette.

Reagents Required

Hydrogen peroxide (30%), sodium sulphide, oxalic acid, hydrochloric acid (2 N), Sodium hydroxide (1 N) or calgon solution (5%) [Calgon is a commercial preparation of mixing 40 gm sodium hexametaphosphate with 10 gm sodium carbonate. This 50 gm calgon when dissolved in water and diluted to a volume of 1 liter gives 5% calgon solution whose pH should be 8.3].

Procedure

Dispersion

40 g air dry processed (i.e. passed through a 2 mm sieve) loamy soil (25 g for clayey soil and 100 g for sandy soil) is taken in 500 mL tall shaped weighed beaker and about 50 mL water is added to make the soil wet. Then about 50 mL hydrogen peroxide is added to the beaker, its content is swirled for few minutes, the beaker is covered with watch glass and left for some time to proceed the reaction in cool. After that the covered beaker is placed on a hot water bath at a temperature of about 65 to 70°C and the content is stirred occasionally with glass rod watching it carefully and removing it from water bath when necessary to prevent frothing over. This process is continued till the reaction completely subsides. In case frothing persists, the process is repeated by adding again 25 to 30 mL of hydrogen peroxide in the beaker and places it on the hot water bath. This process is continued until the whole organic matter is oxidized (i.e. warming produces no further reaction). The beaker is then heated to about 90^0 to 100^0C to remove the excess hydrogen peroxide.

In calcareous soils, 25 mL of diluted (2N) hydrochloric acid is added to the peroxide treated soil and left until effervescence ceases. The suspension is then filtered using Whatman No. 44 filter paper on a buckner funnel connected with a vacuum pump and washed repeatedly with distilled water till it is acid free (when the soil is treated with hydrochloric acid, chloride free is tested with silver nitrate solution).

In red and laterite soils 100 mL of water and a pinch (0.5 to 2 gm each) of oxalic acid and sodium sulphide are added to the beaker before hydrogen peroxide treatment and the beaker is placed on hot water bath for about an hour. The content of the beaker is filtered through a Buchner funnel under vacuum using a Whatman No. 44 filter paper. The soil is repeatedly washed with distilled water till it is free from soluble salts or soluble ions and chloride ions. The soil from the filter paper is then transferred to the same weighed

500 ml beaker with a jet of distilled water and rubber tipped glass rod. Care should be taken so that all fine materials from the surface of the filter paper are removed and minimum quantity of water is used. The content of the beaker is gently evaporated to dryness first on a hot plate and then in an oven at 105^{0}C for about 16 hours. After cooling in a desicator the beaker and the content is weighed and the weight is recorded for subsequent calculation.

25 ml of sodium hydroxide or 20 ml of calgon solution is added to the dry soil and left for over night. The mixture is transferred with water to the cup of a high speed stirrer and the volume is made up to about 400 ml. The mixture is stirred for 5 to 10 minutes. The stirred blades are washed down as they are removed from the suspension.

Fractionation

Separation of course sand

The dispersed soil suspension is transferred to a 1 liter cylinder by pouring it through standard sieve having 0.2 mm diameter which retains coarse sand. The material on the sieve is washed with distilled water along with rubbing gently with rubber tip glass road so that no particles are retained on the sieve. The course sand fraction is transferred to a weighed small container (No.1) by brushing off the sieve surface carefully and the material is dried at 105°C in an oven for about five hours. The container along with sand fraction is cooled in a desicator and weighed.

Separation of silt and clay

The volume of the suspension in the cylinder is made up to one liter by adding distilled water. The cylinder is placed in a temperature controlled chamber. A mark is given on the cylinder at 10 cm depth from the surface of the suspension. The temperature of the suspension is measured and proper of sedimentation for silt plus clay and for clay corresponding to that temperature is noted from table. The content of the cylinder is mixed thoroughly for one minute with the help of a plunger or moved up and down for about 20 times in one minute after putting the stopper. The plunger or stopper is removed and is waited until the swirling motion of the particles has just stopped and steady setting under gravity has just standard. The stop watch is started.

Table: Sedimentation times for soil particles settling through a depth of 10 cm in water (particle density 2.6 gm/cm3)

Temperature (°C)	Settling time with indicated particle diameter					
	2 microns		5 microns		20 microns	
	Hour	Minute	Hour	Minute	Hour	Minute
20	8	0	1	17	4	48
21	7	49	1	15	4	41
22	7	38	1	13	4	35
23	7	27	1	11	4	28
24	7	17	1	10	4	22
25	7	7	1	8	4	16
26	6	57	1	7	4	10
27	6	48	1	5	4	4
28	6	39	1	4	4	0
29	6	31	1	3	4	55
30	6	22	1	1	4	49
31	6	14	1	0	4	44

Observations

a. Weight of the soil taken =..........................gm

b. Weight of container No. 1 =.......................gm

c. Weight of container No. 1 plus oven dry soil after removal of cementing agents and salts =.....................................gm

d. Weight of container No. 2 =.......................gm

e. Weight of container No. 2 plus dry course sand fraction =..........gm

f. Volume of suspension taken for analysis =...........................ml

g. Weight of container No. 3 =...gm

h. Weight of container No. 3 plus dry silt plus clay = ...gm

i. Weight of container No. 4 =...gm

j. Weight of container No. 4 plus dry clay =..........................gm

Calculations

I. Oven dry weight of the soil after removal of cementing agents and salts (c-b) =......gm

II. Weight of course sand (e-d) =...gm

III. Percent course sand (II/I) x 100 =...%

IV. Weight of silt plus clay (h-g) =...gm

V. Percent silt plus clay (IV/I) x (1000/f) x 100 =.......................................%

VI. Weight of clay (j-i) = ...gm

VII. Percent clay (VI/I) x (1000/f) x 100 =...%

VIII. Percent silt (V-VII) =..%

IX. Percent fine sand [100 - (111 + V)] =..%

Result: the percent course sand, fine sand, silt, and clay are .... respectively. By reading

1.7. Determination of Soil Moisture Content (%)

Principle Calculation of the results of soil analysis is done on the basis of "oven dry" soil. The moisture content of air-dry soil is determined prior to soil analysis.

Apparatus Moisture tins or flasks with closely fitted lid, Drying Oven

Procedure 1. Transfer 50 g soil to a tarred moisture box and weigh with 0.001 g accuracy (W_1 or g).

1. Dry over night at 105°C (lid removed)
2. Remove box from Oven, close the lid, cool in a desiccators and weigh (W_2 gram).

Calculation The moisture content (%) in soil is obtained by:

$$\text{Moisture \%} = \frac{W_1 - W_2}{W_2 - \text{tare box}} \times 100$$

The corresponding correction factor for analytical results or amount of sample to be weighed in for analysis is:

$$\frac{10 + \%\ \text{moist}}{\text{Moisture correction factor}} = 100$$

Note : From the view point of soil-plant relationship, it may be argued that for a number of soil parameters it is relevant to express value on a soil volume basis (w/v) rather than on the usual soil weight basis (w/w). Soil weight can easily be converted to volume by using the bulk density. Thus, for instance, a soil with a CEC of 10 me/100 g and a bulk density 1.40 kg/ litre would have a CEC to 14 me/100 ml or 14 centi mol (+)/1.

1.8. Determination of Cation Exchange Capacity (CEC)

Principle

Soil is leached with 1*N* ammonium acetate (pH 7.0). Excess ammonium acetate is removed from the soil with ethanol The exchanged ammonium ions, corresponding to the cation exchange sites on the exchanger are then replaced by potassium ions. Ammonium so desorbed may be distilled to find the ammonium concentration, *vis-a-vis* CEC.

The ammonium acetate extract is examined for exchangeable cations.

A. Preparation of extracts

Reagents

- 1N Ammonium acetate:Dilute approx. 460 ml of acetic acid (glacial) to 2 1. Dilute to approx. 440 ml of ammonia solution (approx. 35% m/m NH_3) to 2 L. Standardize each solution and adjust to 4 M. Mix together 2 litre each 4 M solution and adjust to pH 7.0 with 1 M acetic acid or 1 M ammonia. Dilutc to 8 L.
- Ethanol
- Potassium chloride solution 10% m/v. Add 1 M hydrochloric acid (approx. 2.5 ml/L of solution) until the pH is 2.5.

Extraction

Transfer 5 g air dry soil, ground to pass a 0.5 mm sieve, into a 50 ml borosilicate glass beaker. Add 20 ml of 1N ammonium acetate, stir and allow to stand overnight. Transfer the contents of the beaker to a filter funnel with a 125 mm Whatman No. 44 filter paper and collect the filtrate in a 250 ml volumetric flask. Leach the soil with successive 25 ml 1N ammonium acetate. Collect all the extracts in the flasks containing the original filter and continue leaching until nearly 250 ml have been collected. Dilute to 250 ml with 1N ammonium acetate and retain this solution, solution A, for the determination of calcium, magnesium, manganese, potassium and sodium.

Wash the interior of the funnel, the soil and the filter paper with five volumes of 25 ml ethanol, allow the funnel to drain between washings. Discard the washings.

Leach the ethanol-washed soil on the filter paper with successive 25 ml volumes of potassium chloride solution, allowing the funnel to drain between each addition and collecting the extracts in a 100 ml volumetric flask. Continue leaching until nearly 100 ml have been collected. Dilute to 100 ml with potassium chloride solution and retain this solution, solution B for the determination of cation exchange capacity. Carry out a blank determination.

B. Determination of CEC

Apparatus

Micro-distillation apparatus: Made to the specification of Bremner and Keeney.

Reagents

- Ammonium-nitrogen standard solution, 0.14 mg/mi of nitrogen — Dry ammonium sulphate at 102°C for 1 hour and allow to cool in a desiccator. Dissolve 0.661 g of the dried salt in water and dilute to 1 lit. Store in a refrigerator.
- Boric acid solution — approx. 1% m/v.
- Magnesium hydroxide suspension—Heat magnesium oxide (heavy) at 800°C for 2 h and cool in a desiccator. Suspend 17 g of the ignited oxide in 100 ml of water.
- Methyl red — methylene blue solution — Dissolve 1. 25 g of methyl red and 0.825 g of methylene blue in 1 lit. of ethanol.
- Octan—2-O1.
- Sulphuric acid, 5 mM.

Preparation of apparatus

Steam out the apparatus for 20 miii. Pipette 5 ml of ammonium nitrogen standard solution into the apparatus, add 1 drop Octan—2—ol (as an antifoam), 6 ml of magnesium hydroxide suspension and steam distill the liberated ammonia into 5 ml of boric acid solution. Collect 35-40 ml distillate in 5 min. Add 2-3 drops of methyl red-methylene blue solution and titrate with 5 mM sulphuric acid until the green colour changes to purple. Carry out a blank determination using 5 ml of water in place of the amnionium — nitrogen standard. The titrate of the ammonium-nitrogen standard minus the distillation blank should be 5.0 ml.

Extract Examination

Steam out apparatus for 20 min. Pipette an aliquot (X ml, but not exceeding 50 ml), containing 0.025-1 mg ammonium nitrogen of solution B into the apparatus and continue as described in preparation of apparatus at 'add 1 drop of Octan-2-ol.' and ending at C until the green colour changes to purple. If the titre is less than 0.2 ml, or more than 7.0 ml, repeat the determination using a larger or a smaller, aliquot respectively.

Calculation

$$CEC\ (\frac{me}{100g}\ of\ soil) = \frac{Titre\ value\ of\ soil - Titre\ value\ of\ blank}{Volume\ of\ aliquot\ taken} x\,20$$

1.9. Estimation of Soil Organic Carbon and Organic Matter Content in Soil

Soil organic matter is defined as a whole series of products which range from undecayed plant and animal tissues to fairly stable amorphous brown to black material bearing no trace of the anatomical structure from the material it was derived. In addition, soil organic matter also contains products of microbial synthesis. Thus soil organic matter includes: a) fresh plant and animal residues capable of rapid decomposition b) humus which represents the vast bulk of resistant organic matter and c) inert form of elemental C such as charcoal or coal.

Principle

Soil organic matter is directly correlated to the nitrogen availability to the plant and its determination is often carried out as an index of nitrogen availability as well as soil quality. There are two methods:

(A) Titration method (Walkley and Black 1934)

(B) Colorimetric method (Datta *et al.,* 1962)

In both the method organic matter is oxidized with chromic acid (Potassium dichromate + con. H_2SO_4). Known excess of potassium dichromate is added to the soil sample. After the reaction residual dichromate is titrated back with ferrous sulphate or ferrous ammonium sulphate (redox titration). Carbon is not only the element that is oxidized by chromic acid ($H_2Cr_2O_7$) but H^+ also reduce. Reactions are as follows:

$2H_2Cr_2O_7 + I2H + 6H_2SO_4 = 2Cr_2\ (SO_4)_3 + 14H_2O$

The presence of H^+, therefore, increases the amount of dichromate required for oxidation. However, the carbon present in the carboxylic group does not require dichromate to get oxidized and thus lowers the amount of dichromate required for oxidation.

$R - COOH \rightarrow RH + CO_2$

It is generally accepted that opposite effects of H and carbon (in carboxylic group) balance each other.

Reactions

$4Cr^{6+} + 3 C^0 = 4C^3 + 3C^{4+}$

$2H_2Cr_2O_7 + 3 C^0 + 6H_2SO_4 = 2Cr_2 (SO_4)_3 + CO_2 + 8H_2O$

In the oxidation of organic C the change in valence is from 0 to + 4 (in $C0_2$). The equivalent weight of C in this oxidation reaction is 12/4. Thus 1 ml of 1 $NK_2Cr_2O_7$ will oxidize 3 mg or 0.003 g of C. This explains the factor 0.003. used in calculation.

Colorimetric method the intensity of green colour of chromic acid obtained due to reduction is measured calorimetrically. This intensity is directly proportional to the amount of organic carbon present in the soil.

a) *Walkley and Black method*

Apparatus

1. Conical flask 500 ml (wide mouth)
2. Pipette 1 and 10 ml
3. Safety Pipette 10 ml
4. Automatic burette —50 ml (with 21 container)

Reagents

1. Ortho-phosphoric acid 85% or Sodium fluoride (Na F) solution 2%
3. Sulphuric acid 96%
4. Standard 1 N $K_2Cr_2O_7$: Dissolve 49.04 g of analytical grade $K_2Cr_2O_7$ (dried at 105^{0}C) in distilled water and make the volume dilute it to 1 litre.
5. Standard 0.5 N $FeSO_4(NH_4)_2SO_4.6H_2O$ solution(Mohr'salt). Dissolve 196 g ferrous ammonium sulphate in about 800 ml of distilled water. Add 20 ml concentrated H_2SO_4 and dilute to 11itre.
6. Diphenyl amine indicator—Dissolve 0.5 g reagent grade diphenyl amine in 20 ml water and add 100 ml concentrated H_2SO_4. Store in an amber colour bottle.

Procedure

Measure 1.0 g of the prepared soil sample in 500 ml conical flask. Add 10 ml of 10 N $K_2Cr_2O_7$ solution and 20 ml of conc. H_2SO_4. Mix gently and allow the reaction to complete for 30 minutes. Dilute the reaction mixture with 200 ml water. Add 10 ml H_3PO_4, 1.0 ml of sodium fluroide solution and 2 ml of diphenylamine indicator. Titrate the solution with standard $FeSO_4$ solution to a brilliant green colour. A blank without soil is run simultaneously. Organic carbon in soil is given by the following equation:

$$\text{Organic carbon (\%)} = \frac{10}{5}(F - T)x\frac{0.003x100}{wt\, of\, soil}$$

Since one gram soil is used, this equation simplifies to:

$$\text{Organic carbon (\%)} = 3\frac{(F - T)}{F}$$

Where, F = ml $FeSO_4$ solution required for Blank T = ml $FeSO_4$.solution required for soil sample.

b) Colorimetric method

Apparatus

Colorimeter or spectrophotometer, conical flask, 100 ml centrifuge, pipette, etc.

Reagents

- 1N Potassium dichromate :Dissolve 49.04 of $K_2Cr_2O_7$ in distilled water and make to 1000mL
- Conc. Sulphuric acid (36 N)
- Anhydrous sucrose (AR grade)

Procedure

1. Weight 1 g of 0.2 mm soil sample into a 250 mL conical flask.
2. Add 10 mL of 1N $K_2Cr_2O_7$ and 20 mL of conc. H_2SO_4 while swirling the flask slowly.
3. Swirl a little and keep on an asbestos sheet for 30 minutes and let it attain room temperature.
4. Decant about 15 to 20 mL of the supernatant liquid carefully into a centrifuge tube, retaining soil particles in the flask as far as possible.

5. Centrifuge for about 5 minutes to ensure the complete settling of soil particles, if any,
6. Carefully decant the clear liquid into the colorimeter/spectrophotometer tube and read the colour intensity using red filter/660 nm wave length.
7. Simultaneously run a blank without soil and a series of standards as follows:
    1. Accurately weight 5, 10, 15, 20 and 25 mg anhydrous sucrose crystals into 100 mL conical flasks.
    2. Proceed as above for oxidation of sucrose and colour intensity measurement.
    3. Plot the colorimeter/spectrophotometer readings against quantity of sucrose or carbon.
    4. Find out a factor to be multiplied by sample reading to get organic carbon content (%) of soil, If the standard curve is prepared accurately and repeatedly, the average value of this factor comes to be 0.0042 with little possible variation.

Calculation

Organic carbon (%) = Calorimeter reading X0.0042.

Caution: The centrifuge may be damaged due to accidental spilling of chromic acid during centrifuging the contents. Keeping overnight is suggested as an alternative. Appearance of chromic acid crystals is sometimes observed even before 30 minutes of keeping for oxidation. Dilution of the contents by adding 10 mL of distilled water after 30 minutes reaction followed by keeping overnight is useful if the soil texture is light. For fine textured soils it is, however, necessary to centrifuge.

Estimation of Organic matter = OC (%) x 1.724

Note: Formerly a conversion factor of 1.724 was used, but there are indications that a factor 2 is more appropriate.

Table : Limits of Soil Organic Carbon and Organic Matter (%) as a measure of available N. (Subba Rao 1995)

Range	% OC	% OM
Low	<0.5	<0.86
Medium	0.5-0.75	0.86-1.29
High	>0.75	>1.29

Precautions

1. Add $K_2Cr_2O_7$ very carefully so that it not touch the neck of the flask.
2. Be careful while adding sulphuric acid as it can injure the skin and burn the clothes. The acid should be added through tilt measure only.
3. Read upper meniscus of solution in the burette.
4. If content of the flask turn green with the addition of indicator before titration, repeat the sample with double the volumes of standard potassium dichromate solution and sulphuric acid.
5. Do not place the hot flask on a wet surface as it can break.

1.10. Determination of Available Nitrogen in Soil

Principle

The procedure involves distilling, the soil with alkaline potassium permanganate solution and determining the ammonia liberated. This serves as an index of the available (mineralizable) N status of a soil and was therefore, proposed as a soil test for N by Subbiah and Asija (1956).

Reagents

- 0.32% potassium permanganate $KMnO_4$ solution. Dissolve 3.2 reagent grade $KMnO_4$ in water, make up to 1 litre in volumetric flask and mix well.
- 2.5% NaOH solution: Dissolve 25 g NaOH in distilled water, make the volume to 1litre. Store in a plastic container.
- Liquid paraffin (extra pure).
- 0.02 N(N/50) standard sulphuric acid.
- Boric acid indicator solution (2%): Dissolve 20 g of pure boric acid (H_3BO_3) in about 650 ml of hot water. Transfer the cooled solution to and make up to volume1 litre in volumetric flask. Add 20 ml mixed indicator solution. After mixing the contents of the flask, add approximately 0.005 N NaOH continuously until the colour is reddish purplel (pH 5.0). Then dilute the solution to volume with water and mix it thoroughly.
- Mixed indicator: Dissolve 0.07 g methyl red and 0.1 g bromocresol green in 100 ml of 95% ethanol.

Procedure

- Place 20 g soil in a 800 ml Kjeldahl flask.
- To this, add 20 ml of water and swirl. Then add 1 ml of liquid paraffin and a few glass beads to prevent frothing and bumping, respectively during distillation. Then add 100 ml each of 0.32% $KMnO_4$ and 25% NaOH solution.
- Distill the contents in a Kjeldahl distillation assembly at a steady rate and collect the liberated ammonia in an Erlenmeyer flask (250 ml) containing 20 ml of boric acid solution (with mixed indicator). With the absorption of ammonia, the pink colour of boric acid turns to green. Nearly 100 ml of distillate is to be collected in about 30 minutes.
- Titrate the content with 0.02 $NH_2SO_4$to the original shade (pink). Blank correction (without soil) is to be made for final calculations.

Calculation

$$\text{Mineralisable N kg/ha} = \frac{(S - B) \times N \times 14 \times 2.24^* \times 10^6}{\text{wt. of soil sample (g)} \times 1000}$$

$$= \frac{(S - B) \times N \times 31.36 \times 10^3}{\text{wt. of sample (g)}}$$

Where, S = ml of acid required for sample,

B = ml of acid required for blank, and N=Normality of H_2SO_4=0.02

$$= \frac{(S-B) \times 0.02 \times 31.36 \times 103}{20} = (S-B) \times 31.36$$

*Taking 2.24×10^6 kg ha^{-1}as the weight of 0-15 cm surface soil.

1.11. Determination of Total Nitrogen in Soil

Most of the nitrogen present in soils is associated with organic compounds like proteins, amino acids amino sugars etc. Soil nitrogen also occurs in inorganic forms like ammonium, nitrate, nitrite. As the nitrogen in organic compounds is not easily available, it has to get mineralized to ammonical and nitrate forms before plants can utilize it. So the soil available N which represents the fraction of total nitrogen that can be easily used by the plants comprises ammonical nitrate and the easily oxidisable organic form.

Principle

For determination of total N in soil or plant by Kjeldahl method, organic N in the sample under analysis is converted to NH_4—N by digestion with concentrated H_2SO_4containing substances (K_2SO_4 or Na_2SO_4) that promote this conversion. The NH_4—N in the digest is determined from the amount of NH_3 liberated by distillation of the digest with an alkali (NaOH). Ammonia absorbed in the boric acid containing mixed indicator is determined with a standard acid.

Reactions

(i) Digestion

$(C_6H_{10}O_5) + 2n\ H_2SO_4 = 6n\ CO_2 + 7n\ H_2O + 2n\ SO_2$

$2[R\text{-}COONH_2] + 2H_2SO_4 + 5H_2 = 2(NH_4)_2SO_4 + 2[R\text{-}\ COOH]$

(ii) *Distillation*

$2\ (NH_4)2\ SO_4 + 4\ NaOH = 2\ Na_2SO_4 + 4H_2O + 2NH_3$

$2NH_3 + 2H_2O = 2NH_4OH$

$NH_4OH + H_3BO_3 = NH_4[B\ (OH)_4]$

(iii) *Titration*

$2NH_4\ [B\ (OH)_4] + H_2SO_4 = (NH_4)2\ SO_4 + 2H_3BO_3 + 2H_2O$

Apparatus needed

1. Kjeldahl digestion assembly
2. Ammonia distillation assembly
3. Conical flask 250 ml
4. Pipette 25 ml
5. Kjeldahl flasks 500/800 ml
6. Automatic burette 50 ml with 2 liters container.

Reagents

1. Conc. H_2SO_4
2. Granulated Zinc
3. Standard $N/10H_2SO_4$
4. Sodium thiosulphate ($Na_2S_2O_3$)
5. Salicylic acid ($C_6H_4OH\ COOH$)
6. Sodium sulphide (Na_2S)

7. Digestion accelerator and catalyst mixture: Mix by grinding in a mortar approx. 100 g anhydrous Na_2SO_4 or K_2SO_4, 10 g $CuSO_4 . 5H_2O$ and 1 g Selenium powder.
8. 40% NaOH: Use commercial grade alkali, the impurities to be removed.
9. Mixed indicator: Dissolve 05 g of Bromocresol green and 01 g of methyl red into 100 ml Ethanol.
10. Boric acid indicator solution (4%): Place 40 g pure boric acid (H_3BO_3) in litre flask and add about 900 ml of distilled water. Heat and swirl the flask until the H_3BO_3 is dissolved. Cool the solution, and add 20 ml of mixed indicator solution. Then add 0.1 N sodium hydroxide solution cautiously until the solution assumes a reddish purple tint (pH5.0), and make the volume to litre.

Procedure

The procedure for determination of total N in soils involves the three main steps.

1. Digestion; 2. Distillation; and 3. Titration.

Digestion

(i) Take 5 g soil sample (0.5 g of muck or peat or 10-20 g of. sandy soil) in 500 ml Kjeldahl flask.

(ii) Add 25 ml conc. H_2SO_4 containing 1 g salicylic acid — Salicylic acid is added to include Nitrates & Nitrites of soil.

(iii) Keep it for few minutes after shaking and add 5 g sodium thiosulphate ($Na_2S_2O_3$). Heat the material for 5 minutes and cool.

(iv) Add 20 ml of water, 1 g digestion accelerator and catalyst mixture and digest the contents till light blue colour appears.

Distillation

(i) Cool the contents and add about 50 ml water to the digestion flask and swirl the flask for 2 minutes. Take the supernatant liquid into distillation flask. Repeat this process atleast four times.

(ii) Add 10 ml Na_2S solution to the distillation flask and then add 135 ml of 40% NaOH slowly from the side of the flask. Add 2 pieces of Zn and distill NH_3 into 25 ml boric acid indicator solution kept in a receiver flask.

(iii) When no more NH_3 is received-test with blue litmus — stop the distillation and proceed for titration.

Titration

The unreacted boric acid solution in the receiver flask is back titrated with standard H_2SO_4 The disappearance of blue colour indicates the end point.

Note : A blank without the soil sample is to be carried out.

Calculation

$$\%N = \frac{xn - x0}{w} xNx1.4xM$$

Where,

x^n = ml H_2SO_4 required for sample

x° = ml H_2SO_4required for blank

W = Sample weight g

N = Normality of H_2SO_4

1.4 = 14 x 10^{-3} x 100% (14 = atomic weight of nitrogen)

M = moisture correction factor.

1.12. Determination of Available Phosphorus in Soil

Phosphorus occurs in soils, both in organic and inorganic forms. Organic fraction is present in humus and other compounds like phytins and nucleic acids. Inorganic fraction occurs in combination with calcium, iron, aluminium and other elements. The content of inorganic phosphorus (Ca, Fe and Al-phosphate) in soils is higher than that of organic phosphorus except in organic soils. Plants absorb P from soil as $H_2PO_4^-$ and HPO_4^{--}.Most commonly used methods for the estimation of available P is soil are:

(A) The Olsen's Method or Sodium bicarbonate extractable P : The sample is extracted with sodium bicarbonate solution (0.5 M $NaHCO_3$) of 8.5 pH. This method is suitable for neutral to alkaline soils (Olsen et al, 1954)

(B) Bray and Kurtz No. 1 Method : The sample is extracted with combination of Ammonium fluoride and Hydrochloric acid (0.03 M NH_4 F + 0.25 N HC1) (Bray and Kurtz 1945). This method is suitable for acidic soil.

Principle

After extraction of P from soil, Phosphate in the extract is measured by the reaction of phosphate with molybdate in an acid medium to form molybdophosphoric acid. Molybdophosphoric acid is then reduced to a blue coloured complex (reduced phosphomolybdenum blue) through reaction with ascorbic acid. Absorbance reading are taken at 730 nm wavelength using a spectrophotometer. A standard curve plotted from absorbance reading of standards is used to deduce phosphate concentration of sample.

$H_3PO_4 + 12H_2MoO_4 \rightarrow H_3\,P\,(Mo_3O_{10})_4 + 12H_2O$

Molybdophosphoric acid.

A) ***Olsen's Method***

Apparatus

- Spectrophotometer (with 10 mm cuvette).
- Polythene shaking bottles 250 ml.
- Reciprocating shaking machine.

Reagents

- Sodium bicarbonate solution (0.5 M) pH 8.5 (extracting solution): Dissolve 42 g $NaHCO_3$ in 500 ml water and make to 1 liter Adjust the pH to 8.5 by adding NaOH 1 M (4 g/100 ml). (Note : Check and read just the pH after storage)
- Sulphuric acid (4 M) Slowly add 56 ml concentrated H_2SO_4 to 150 ml water under constant stirring. After cooling make to 250 ml with water.
- Ammonium molybdate solution (4%) Dissolve 4 g of Ammonium molybdate AR grade, $(NH_4)_6MO_7O_{24}$.$4H_2O$ in water and make to 100 ml. Store in polythene or Pyrex bottle.
- Potassium antimonytartrate solution, 0.275% (1000 ppm Sb) : Dissolve 0.275 g $KSbOC_4H_4O_6$in water and make to 100 ml.
- Ascorbic acid solution, 1.75% : Dissolve 1.75 g ascorbic acid in water and make to 100 ml Prepare fresh daily.
- Mixed reagent (Prepare fresh daily) : Successively add with a measuring

cylinder to a 500 ml polythene or Pyrex bottle and homogenize after addition:

- 50 ml of 4 MH_2SO_4
- 15 ml of ammonium molybdate solution
- 30 ml of ascorbic acid solution
- 5 ml of KSb-tartrate solution
- 200 ml of water

Standard phosphate solution, 100 ppm P

- To make 100 ppm P solution dissolve 0.4390 g KH_2PO_4 in water in a 1 liter volumetric flask and make to volume.
- Standard phosphate solution, 4 ppm P : Pipette 10 ml of the 100 ppm P standard solution into a 250 ml volumetric flask and make to volume with extracting solution.
- Standard series : Pipette into 100 ml volumetric flasks 0-10-20-30-40-50 ml of 4 ppm P standard solution. Make to volume with extracting solution. The standard series is then 0-0.4-0.8-7.2-1.5-2.0 ppm P.

Procedure

1. Weigh 5 g fine soil (accuracy 0.001 g) into 250 ml polythene shaking bottle. Include blank and a reference sample.
2. Add 100 ml of the extracting solution
3. Shake for 30 minutes.
4. Filter through Whatman No. 1 filter paper.
5. Pipette into (short) test tubes 3 ml of the standard series, the blank and the S extracts.
6. Slowly add 3 ml of mixed reagent by pipette and swirl (CO_2evolution). Note: mixed reagent contains excess H_2SO_4 to neutralize the $NaHCO_3$bringing the pH about 5.0.
7. Allow the solutions to stand for at least 1 hour to develop maximum blue colour.
8. Measure absorbance on spectrophotometer at 882 or 720 nm.

Calculation

Plot a calibration graph of absorbance against P concentration:

P(ppm or mg/kgsoil) =(x—b) x v/s x m

Where

x = ppm P in sample extract

b = asblank

v = volume Of extracting solution taken (100 ml)

s = sample weight in gram (5 g)

m = moisture correction factor

Olsen's P (kgha^{-1}) = 8.96 X quantity of P in ppm or mg kg^{-1}

Conversion factor: P_2O_5 =2.29 x P

B) Bray's and Kurtz No. 1 method

Apparatus

- Spectrophotometer (with 10 mm cuvette)
- Mechanical Shaker

Reagents

- Ammonium fluoride solution (1 M) Dissolve 3.7 g NH_4F in water and make to 100 ml(store in a polythene bottle).
- Hydrochloric acid (0.5 M) :Dilute 8.3 ml HCl 6 M (or 4.3 ml conc. HCl) to 100 ml with water.
- Extracting solution Bray 1 (0.03 M NH_4F and 0.025 M HCl). :Add 15 ml NH_4F 1 M and 25 ml HCl 0.5 M to 460 ml water. (This solution can be kept in glass for over a year).
- Sulphuric acid (2.5 M) : Slowly add 35 ml concentrated H_2SO_4 to 150 ml water under constant stirring. After cooling make to 250 ml with water.
- Ammonium molybdate solution, (4%) : Dissolve 4 g of $(NH_4)_6 Mo_7O_{24}$. $4H_2O$ in water and make to 100 ml. Store in polythene or Pyrex bottle in the dark.
- Potassium antimony tartrate solution, (0.275%) (1000 ppm Sb) :Dissolve 0.275 g KSb $OC_4H_4O_6$ in water and make to 100 ml.

- Ascorbic acid solution, (1 .75%) :Dissolve 1.75 g ascorbic acid in water and make to 100 ml. Prepare fresh daily.
- Mixed reagent (Prepare fresh daily) : Successively add with a measuring cylinder to a 500 ml polythene or Pyrex bottle and homogenize after each addition:
  - 50ml of 2.5 MH_2SO_4
  - 15 ml of NH_4-molybdate solution
  - 30 ml of ascorbic acid solution
  - 5 ml of KSb-tartrate solution
  - 200 ml of water.

Boric acid solution, (1%): Dissolve 1 g H_3BO_3 in 100 ml water.

Standard phosphate solution, 100 ppm P

- Dissolve 0. 4390 g KH_2PO_4 in water in a 1 litre. volumetric flask and make to volume. The strength of the solution will be 100 ppm P.
- Pipette 25 ml of the 100 ppm P standard solution into a 250 ml volumetric flask and make up to volume with water to prepare 10 ppm P solution.
- Standard series: Pipette into 100 ml volumetric flasks 0-10-20-30-40-50-60-70-80-90-100 ml of the 10 ppm P standard solution. Make to volume with distilled water. The standard series is then 0-1-2-3-4-5-6-7-8-9-10 ppm P.

Procedure

1. Weigh 2 g fine soil (accuracy 0.01 g) into a wide month (50 ml) or shaking bottle. Include two blank and a reference sample.
2. Add 14.0 ml of extracting solution Bray 1.
3. Shake exactly for 1 minute by hand and then immediately filter through a hardened filter.

Note: In case the filtrate is turbid filter again through the same filter. Filtration procedure not to exceed 10 min.

4. Pipette into (short) test tubes 1 ml of the standard series, the blanks and the sample extract S, 2 ml boric acid and 3 ml of mixed reagent. Homogenize.

5. Allow solutions to stand for at least 1 hour for the blue colour to develop its maxim (see remarks below).
6. Measure absorbance on spectrophotometer at 882 or 720 mm.

Calculation

Plot a calibration graph of absorbance against P concentration

P(ppm or mg/kgsoil) = (x-b)x v/s x m

where

x = ppm P in sample extract

b = ditto inbank

s = sample weight in gram, 2g

v = amount of extracting solution taken, 14 ml

M= moisture correction factor

Conversion factor: P_2O_5 =2.29 x P

Remarks : With the acid molybdate solution phosphate forms phospho-molybdenic ac which is reduced to phospho-molybdenic-blue with ascorbic acid. The antimony accelerates the development of the blue colour and stabilizes it for up to 24 hours. With this method interference cf Si is not to be expected. If such an interference still occurs (blue coloured standard) then repeat procsdure using distilled water (Naturally, distilled water may be used from the start).

Available P ($Kgha^{-1}$) = $\frac{C \times v \times 2.24 \times 10^6}{S \times 1} = \frac{C \times v \times 2.24}{S \times 1}$

Where,

C = Quantity of P in ug or ppm read on x axis against a sample reading-ditto blank reading

v = volume of extracting reagent used (ml)

I = volume of aliquot used for colour development (ml)

S = wt. of soil sample taken (g)

Thus Brays P($kg.ha^{-1}$) = CX4.48 and Olsen's P ($kg.ha^{-1}$) = Cx896

(a) Basis for classifying soils as low medium and high in available P : Olsen et al. (1954) used wheat, oats, alfalfa and cotton as test crops in their studies and 0.5 M$NaHCO_3$Soluble P values were classed as follows : 5 ppm P-soils low in available P, responsive top application; 5-10 ppm P-soils medium in available P, may respond to P application. > 10 ppm P-soils high in available P, response to P application is not likely. The limits for low, medium and high available P used by soil testing laboratories in India are below 10, 10-24 .6 and above 24. 6 kg P ha^{-1}, respectively (Subba Rao, 1995). These are general ratings for all crops.

(b) Effect of temperature of extracting solution: Temperature of the extracting solution and shaking speed are the possible sources of error. The temperature used by Olsen et al.(1954) was 25°C ± 1°C. For soils testing between 5 and 40 ppm P, each degree rise in temperature between 20°C and 30°C led to an increase in P extraction by 0.43 ppm.

(c) Range of analytical procedure : It works in the range of 0.05 to 1 ppm P. Concentrations of 2000 ppm or more of CV, 15 ppm of Fe^{+3}, 1000 ppm of $A1^{3+}$, 200 ppm of Ca^{2+} 600 ppm of SO_4^{2-}, 50 ppm of $C1O_4$ or 25 ppm of NO_3^- do not interfere.

(d) Functions of reagents: 1. Ammonium molybdate : It forms yellow coloured molybdophosphate complex with P in soil extract.

1.13. Determination of Total Phosphorus in Soil

Phosphorous occurs in soils both in organic and inorganic forms. Organic fraction is present in humus and other compounds like phytins and nucleic acids. Inorganic fraction occurs in combination with calcium, iron, aluminium and other elements. The content of inorganic phosphorus (Ca, Fe and Al phosphates) in soils is higher than that of organic phosphorus except in organic soils. Plants absorb P from the soil as $H_2PO_4^-$ and HPO_4^{-2} ions. Since the contents of these forms of Pin soil solution is very small, it must be replenished during the growing the season of the crop. To accomplish that, it is important to know the available P status of the soil.

For the estimation of available P, extractants such as water, dilute acid, alkalies and salt solutions have been used. In neutral to alkaline soils, the available P is extracted with 0.5M $NaHCO_3$ (pH 8.5) as detailed by Olsen et al. (1954).

1.13.1 Total Phosphorus in Soil

Principle

Total P in soil is determined by two methods

1. Fusion of soil with Sodium Carbonate (Na_2CO_3).

2. Digestion of soil with perchloric acid ($HClO_4$) some worker says that sodium carbonate fusion is better method while, Jackson (1958) observed the similar results from both methods for most of soils and rocks. Olsen and Sommers (1982) gave the procedure of digestion of soil with perchloric acid ($HClO_4$) which is convenient for agronomy soil fertility laboratories.

Apparatus

- Spectrophotometer (with 10 mm cuvettle)
- Volumetric flask; 250 and 1000 ml
- Pipette etc.

Reagents

1. Percloric Acid ($HC1O_4$) 60%
2. Ammonium Paramolybdate Solution
   - Dissolve 2.5 g of ammonium molybdate $[(NH_4)_6 \cdot MO_7O_{24} \cdot 4H_2O]$ in 400 ml of distilled water.
   - Dissolve 1.25 g of ammonium metavanadate (NH_4VO_3) in 300 ml of boiling distilled water in 1 litre volumetric flask.
   - Cool the solution and add 250 ml of concentrated nitric acid (HNO_3) & cool the solution to room temperature.
   - Pore the ammonium molybdate solution into ammonium metavanadate + nitric acid solution and make up the volume 1 litre with distilled water.
3. Standard Phosphate Solution
   - Dissolve 0.4390 g of oven-dried (at 60°C for 1 hr. and cooled in desiccator) potassium dihydrogen orthophosphate (KH_2PO_4) in distilled water in 1 litre volumetric flask and make to volume.
   - One ml, of this solution contains 100 μg P or the strength of this solution will be 100 ppm.
   - Pipette 25 ml of 100 ppm standard solution into a 250 ml volumetric flask and making to volume with distilled water to prepare 10 ppm P solution.

- Standard series : pipette into 100 ml volumetric flask 0-20-40-60-80-100 ml of 100 ppm P. Standard solution. Make volume with distilled water. Standard series is then 0-2040-60-80-100 ppm P.

Procedure

1. Take 2 g soil sample and 30 ml of $HC1O_4$ (60%) in 250 ml volumetric or Erlenmeyer flask.
2. Digest in mixture of a temperature a few degree below the boiling point on a hot plate in fume hood chamber until the dark colour due to organic matter disappears. Then continue heating at the boiling temperature for nearly 25 minutes longer. At this stage heavy white fumes of $HC1O_4$ appears, and the insoluble mineral becomes like white sand. Cool the mixture and add distilled water to obtain a volume of 250 nil id mix the contents. When solid material settle down then take aliquot.
3. Pipette aliquot that contains 0.05 to. 10 mg of P of the solution into 50 ml volumetric flask. Add 10 ml of vanadomolydate reagent and dilute the solution to 50 ml with distilled water.
4. Measure the optical density at 600 nm wavelength.
5. Prepare a reagent blank and subtract this optical density reading from sample readings.

Calculation

P in soil (ppm) = ppm P raead from standard curve x dilution factor

P in kg/ha=P in ppm x2.24

1.14. Determination of Available Potassium in Soil

About 90-98% of total potassium in soil is present in minerals, such as potash feldspars, muscovite and biotite micas. One to 10% of total K is also found in secondary clay minerals like illites vermiculites and chlorites. Potassium in these expanding type minerals forms an integral part of the crystal lattice and cannot be replaced by ordinary exchange methods. It is therefore, referred to as non-exchangeable K.

The term available potassium (K) conventionally refers to exchangeable + water soluble K. The exchangeable K constitutes the major portion of available K except in saline and saline sodic soil. Available K or exchangeable K along with Ca and Mg are usually determined in neutral Normal ammonium acetate

(N NH_4OAC) extract of soil. The extraction is carried out by shaking followed by filtration or centrifugation. The K is estimated using flame photometer. In soils with appreciable amount of soluble K, Ca and Mg, these cations are estimated in a saturation extract (Jackson, 1958) and deducted from N NH_4OAC extractable K, Ca and Mg to obtain respective exchangeable cations.

Apparatus

- Flame Photometer
- Mechanical Shaker
- pH meter
- Conical flask etc.

Reagents

- **Potassium chloride standard solution:** Make a stock solution of 1000 ppm K by dissolving 1.908 g of AR grade potassium chloride (dried at 60°C for 1 hour) in distilled water and diluting to 1L. Prepare 100 ppm K standard by diluting 100 ml of 1000 ppm stock solution to 1L with extacting solution.
- **Ammonium acetate: 1.0 N, pH 7.0 :** To 700 ml of distilled water, add 57 ml of 99.5% glacial acetic acid (CH_3COOH) and then 69 ml of concentrated ammonium hydroxide (NH_4OH). Dilute to volume 900 ml and adjust pH to 7.0 by the addition of more of 3N NH_4OH or 3N CH_3COOH and make up to 1L. Store in a Pyrex bottle. Alternatively dissolve 154 g of ammonium acetate (CH_3COONH_4) in water and dilute to 1.8 lt. Mix thoroughly. Adjust pH to 7.0 with dilute NH_4OH or HOAC as required and make to 2l.
- **Preparation of Standard curve :** Pipette 0,5, 10, 10, 15, and 20 ml of 100 ppm solution into 100 ml volumetric flasks and bring the volume to mark with extracting solution. The solution contain 0, 5, 10, 15 and 20 ppm K respectively.

Procedure

Shaking and filtration can be done by the method of Hanway and Heidel (1952) or shaking and centrifugation as described by Knudsen et al. (1982).

(a) Shaking and filtration (Hanway and Heidel, 1952)

- Weigh 5 g of soil sample in 100 mL conical flask.

- Add 25 ml of neutral 1 N ammonium acetate solution and shake for 5-10 minutes.
- Filter through Whatman No. 1 filter paper.
- Measure K concentration in filtrate using Flame photometer.

(b) Shaking and centrifugation (Knudsen et al., 1982).

- Place 10 g of <2 mm air dried soil (or use 5 g, if the soil contains greater than 500 ppm (K) in a 50 ml centrifuge tube.
- Add 25 ml NH_4OAC stopper and shake the tube for 10 min.
- Centrifuge the tube at 2000 rpm for 10 mm. until the supernatant liquid clear. `
- Decant the supernatant liquid into a 100 ml volumetric flask.
- Make three additional extractions in the same manner. Dilute the combined extracts to 100 ml with NH_4OAC.
- Mix the solution and determine K in the extract in the flame photometer using K filter after necessary setting and calibration of the instrument (use 0 and 20 ppm working K concentration).
- Read similarly the different concentrations of K in flame photometer and obtain the standard curve by plotting the reading against the different concentrations of K.

Calculation

Available K (kg/ha)

Where, C stands for the concentration of potassium in the sample obtained. on X-axis, against the reading.

K_2O=K x 1.2

Note

- The filtrate should be clear in order to avoid choking of capillary tube of the flame photometer which occurs very frequently.
- During the operation of flame photometer, a constant air pressure and steady flow of gas (LPG) and air combination is absolutely necessary for precise estimation. An appropriate combination is achieved by fixing the air pressure around 0.45 psi and then adjusting the gas supply so as to get a non-luminous blue flame.

- Fans and coolers should be switched off so that flame is not affected.
- Ammonia bottle should be cooled before opening.
- Potassium standards should be prepared fresh after every 2-3 weeks.

1.15. Determination of Total Potassium in Soil

Potassium in soil exists as water soluble, exchangeable, non-exchangeable (fixed) and lattic-K. Potash in Indian soil ranges from 0.05—3.5% out of which 95% part is present in complexed form, 1-10% part in relatively non-available form, and 2% in available form. Available potassium includes only exchangeable and water soluble form, whereas total potassium is considered all kinds of potassium present in soil.

Principle

The most widely used digestion technique for determination of total potassium in soils and minerals have been combinations of Hydrofluoric acid (HF) and either H_2SO_4 or perchloric acid ($HClO_4$). HF decomposes silicates by the reaction of F^- with Si to form SiF4 which is volatile when heated in the presence of strong acids. The organic matter is decomposed by the $HClO_4$, which becomes a strong oxidizing agent when hot and concentrated. Organic soils can be pretreated with HNO_3 and $HClO_4$ to oxidize the relatively large amounts of organic matter present and to avoid danger of an explosion during the process of SIF_4 removal. The $HClO_4$ also effectively removes the excess HF from the sample. The $HClO_4$ is preferred to H_2SO_4 because it is more effective in removal of organic matter and also produces less interference in flame photometric procedures.

Apparatus

- Flame photometer
- Crucible
- Sand bath
- Meker burner
- Hot plate
- Volumetric flask 10, 100 mL

Reagents

- Hydrofluoric acid (HF), 48%
- Perchloric acid ($HClO_4$), 70 to 72%

- Nitric acid (HNO_3), 70%
- Hydrochloric acid (HCl), 6 N.

Procedure

- Place a 0.1 g sample of finely ground soil in a 30 ml crucible.
- Wet the soil with a few drops of water and add 5 ml of HF and 0.5 ml of $HC1O_{4.}$
- Heat the soil acid mixture on a hot plate until fumes of $HC1O_4$ appear.
- Cool the crucible and then add 5 ml of HF.
- Place the crucible in a sand bath and cover about nine-tenths of the crucible top with a platinum lid.
- Heat the crucible to 200-225°C and evaporate the contents to dryness.
- Cool the crucible and add 2 ml of water and a few drops of $HC1O_4$.
- Replace the crucible in the sand bath, and evaporate the content to dryness. If organic matter stains are still present on the sides or lid of the crucible, direct the flame of a Meker burner on to the sides and lid until the organic matter is oxidized. A faint red heat is sufficient.
- Remove the crucible, and when it is cool, add 5 ml of 6 N HC1 and about 5 ml of water.
- Heat the crucible on a hot plate or over a burner until the solution boils gently.
- If the sample does not dissolve completely, evaporate the solution to dryness and repeat the procedure, starting with 5 ml and 0.5 ml of $HC1O_4$.
- Filter the sample into 100 ml volumetric flask, wash the filter paper with water and dilute the solution to volume and read the concentration on a flame photometer.

Calculation

$$\text{Total K in soil (kg/ha)} = \frac{C\,x\,volume\,of\,extract}{weight\,of\,soil\,taken}\,x\,2.24$$

Where, C stands for the concentration of K in the sample obtained on x axis of standard curve against reading.

K_2O=K x 1.2

1.16. Determination of Exchangeable Calcium and Magnesium in Soil

Principle

Exchangeable Ca and Mg can be determined in ammonium acetate extract of soils (obtained as discussed under K) by direct titration with EDTA. The presence of ammonium acetate does not interfere with the titration if this is carried out by the procedure suggested below (Hesse, 1971). The amount of organic matter dissolved is usually too small to affect the colour change of the indicator. The procedure permits the determination of Ca and Mg in the same solution.

Reagents

- Standard calcium solution : weight 0.5005 g of $CaCO_3$ dried at 150°C into a 1lt volumetric flask. Add 200 ml of water and then 150 ml of 1 M HCl slowly and with shaking. Make up to 1lt. This solution is 0. 01 N Ca.
- EDTA solution : Dissolve 20 g of EDTA (di-sodium salt) in water and make up to 1 liter.
- Sodium hydroxide aqueous solution: 10% w/v.
- Buffer solution (pH 10) : Dissolve 67.5 g NH_4C1 in 400 ml water; add 570 ml of conc. NH_4OH and dilute to 1lt.
- Hydroxylamine-hydrochloride aqueous solution : 5% w/v. Prepare fresh each week. Potassium hexacyanoferrate (II) aqueous solution (Potassium ferrocyanide) 4% w/v.
- Potassium cyanide aqueous solution (EBT): 1% w/v.
- Trithanolamine : commercial reagent
- Calcon solution: Dissolve 0.2 g reagent in 50 ml methanol. Prepare fresh every 2 weeks.
- Eriochrome Black T solution: Dissolve 0.2 g reagent in 50 ml methanol, prepare fresh every two weeks.

Procedure

Standardization of EDTA solution : Pipette 5 ml of standard calcium solution into a graduated, tall form, 100 ml beaker. Dilute to 10 ml and add 15 ml of ammonium chloride-hydroxide buffer solution. Add 10 drops each of potassium cyanide, hydroxylamine- hydrochloride, potassium hexa cyanoferrate, triethanolamine and EBT solutions. Prepare a blank solution in exactly the same manner, taking 5 ml of water instead of calcium solution. Usually the blank solution .will be blue in colour, but if not, it should be titrated with EDTA solution until blue, and the blank titre value noted. Keep the blue blank solution alongside the standard calcium solution and titrate the standard with EDTA solution, stirring all the time to a permanent blue colour matching the blank. Dilute the blank with water now and again equalize the volumes of the two solutions as the titration proceeds.

Repeat the standardization using calcon as indicator. In this case add to the diluted calcium solution 10 drops each of potassium cyanide, hydroxylamine hydrochloride, and triethanolamine solutions. Then add 2.5 ml NaOH solution and 1 ml of calcon solution. Prepare a reagent blank and titrate both solutions with EDTA solution until blue.

Determination of C in NH_4OAC extract : Pipette an aliquot of the extract containing up to 3 mg Ca into a 100 ml tall-form beaker and dilute to 10 ml. For soils containing less than 10 me Ca/100 g or solution containing less than 1 mg Ca/10 ml, it is necessary to take larger aliquots in a 250 ml beaker. Add 10 drops each of potassium cyanide, hydroxylamine-hydrochloride and triethanolamine solutions. Add. 2.5 ml NaOH solution and 1 ml of calcon solution. Titrate with EDTA until blue.

Determination of Ca + Mg : Pipette an aliquot containing up to 3 mg (Ca + Mg) into a beaker. Dilute to 10 ml, add 15 ml of ammonium chloride-hydroxide buffer solution. and then 10 drops each of potassium cyanide, hydroxylantine-hydrochioride, potassium hexacyanoferrate (II) and triethanolainine solutions while gently warming the solution on the magnetic stirrer. When all reagents have been added, continue working for 3 minutes, cool and add 10 drops of EBT solution. Titrate with EDTA.

Determination of Mg : For most purposes Mg is calculated from the difference between the (Ca + Mg). and the Ca determinations.

Calculations

$$\text{Ca or (Ca + Mg), meq/l} = \frac{\text{T X normality of EDTA 1000}}{\text{Aliquot (ml) taken}}$$

Where T = Volume in (ml) of standard EDTA used in titration.

$$\text{Ca or (Ca + Mg), meq 100 g soil} = \frac{100}{soil\ wt\ (g)} x \frac{extract\ volume}{1000} x\, meq\, Ca\, or\ (Ca + Mg)/lt$$

1.17. Estimation of Available Sulphur in Soil

For estimating plant available sulphur in soils, heat soluble sulphur and 0.15%

$CaCl_2$ extractable S methods (Williams and Steinbergs, 1959) is usually recommended.

Hot soluble Sulphur

Principle

Inorganic sulphate and a portion of organically held SO_4^{2-} are mobilized by special heat treatments and extraction with 1.0 per cent sodium chloride. Sulphate S is converted into $BaSO_4$ suspension using $BaC1_2$ crystals and a special conditioning agent. The resulting turbidity is determined by spectrophotometer.

Apparatus

- Spectrophotometer or Colorimeter or Autoanalyzer
- Hot air Oven, Water bath, centrifuge with 50 ml tubes, Shaker
- Silica Basin, Erlenmeyer flask 150 ml, etc.

Reagents

- 1% sodium chloride (NaCI) : Dissolve 10 g of NaCI in 700 ml of distilled water and make volume to 1 lit.
- Barium chloride dihydrate : $BaC1_2$. $2H_2O$ crystals, 20-30 mesh.
- Conditioning reagent: Dissolve 75 g NaC1 in 275 ml of distilled water in a 500 ml volumetric flask, stirring with magnetic stirring bar, add 30 ml conc. HC1, 100 ml of absolute ethanol and 50 ml of glycerol Rinse glycerol into, flask. Continue Stirring until NaC1 dissolves. Remove stirring bar and make to volume with distilled water.
- Standard sulphate solution: Dissolve 0.5437 g K_2SO_4 in 1 liter water. This contains 100 mg S/lit. Dilute 10 times to obtain 10 ppm S solution. Transfer 0,5, 10, 15 and 20 ml (containing 0, 50, 100, 150 and 200 mg SO_4-S) to Erlenmeyer flask of 150 ml, bringing' the volume with distilled water to 30 ml.

Procedure

- Weigh 5 g soil into a silica basin and add 20 ml distilled water.
- Place the basin on a gently boiling water bath to dryness.
- Then, heat in a hot-air oven at 102°C for 60 minutes.

- After cooling, transfer the soil to a 50 ml centrifuge tube and extract with 33 ml of 1.0% NaCl.
- 25 ml of aliquot is pipetted into a silica basin and evaporated to dryness in 2 ml of 3% H_2O_2. This is done to remove interfering organic matter. The basin is then heated in a hot air oven at 102°C for 60 minutes to ensure the removal of excess H_2O_2. After cooling, residue is taken up in 25 ml water, transferred to a centrifuge tube and centrifuged to remove suspended matter. Sulphur is then determined by taking a suitable aliquot.
- Pipette l0 or 20 ml extract in a 150 ml Erlenmeyer flask and add 20 or 10 ml of H_2O, thus bringing the volume to 30 ml.
- Add 2.5 ml of stabilizing solution and 0.2 to 0.3 g $BaC1_2$ crystals by a spatula to standards and extracts.
- Shake the flasks for 1 minute each to a constant rate.
- After 1-3 minutes, measure the turbidity in a spectrophotometer at 440 nm. The turbidity remains constant during 3-10 minutes.

Calculation

$$SO_4 - S \text{ mg/kg} = \frac{\text{S (mg/1) in aliquot x Volume of NaC1 taken (ml)}}{\text{Volume of aliquot (ml) x wt. of Oven dry soil (g)}}$$

Calcium Chloride ($CaCl_2$) Extractable Sulphur

Principle

The 0.15% $CaCl_2$ extractable S method is fairly quick. Since air drying of soil samples normally increases the amount of S extractable by $CaC1_2$ it is suggested to use (whenever possible) field moist samples.

Reagents

- $CaC1_2$ dihydrate ($CaCl_2$. $2H_2O$) 0.15%: Dissolve 1.5 g $CaC1_2$, $2H_2O$ in about 600 ml water and make up to 11 with distilled water.
- $BaCl_2$ dihydrate ($BaCl_2$, $2H_2O$): Crystals 20-30 mesh (crystal passing through 1 mm sieve).
- Conditioning reagent: Dissolve 75 g NaC1 in 275 ml distilled water in a 500 ml vol. flask. Stir well, add 30 ml conc. HCl, 100 ml absolute ethanol and 50 ml glycerol. Rinse it. Make up the volume with distilled water.

- Standard sulphate solution Dissolve 0.5434 K_2SO_4 (AR) in distilled water and dilute to 1litre. This solution is 100 ppm or 100 mg S/l. Dilute 10 times to obtain 10 ppm solution.

Procedure

- Standard curve : Transfer 0, 5, 10, 15 and 20 ml of 10 ppm standard sulphate solution in Erlenmeyer flask and make up to 150 ml by adding distilled water. Keep the flasks ready for making standard curve.
- Sample preparation : Transfer 5 g soil into an 150 ml Erlenmeyer flask.
- Add 25 ml 0.15% $CaCl_2$ and shake well in a shaker for 30 minutes.
- Add 2.5 ml of stabilizing solution and 0.2 to 0.3 g $BaCl_2$ crystals by a spatula and shake the flask for 1 minute.
- Repeat step 4 with standards.
- After 1 to 3 minutes measure the turbidity in a colorimeter using a blue filter at 440 nm on a spectrophotometer. The turbidity remains constant for 10 minutes.
- Draw a standard curve and record the readings for the samples.

Calculations

ppm SO_4—S in soil = Reading for the standard curve x 25/5

SO_4 — Skg ha^{-1} = ppm SO_4 — S x 2.24

(*taking 2.24 million kg ha^{-1} weight of 0-15 cm soil)

(25/5 = ml of 0.15% $CaCl_2$/wt of soil sample)

1.18. Determination of Total Sulphur in Soil

Sulphur occurs in soil in both organic and inorganic forms, but only a fraction of it is available to plant by the direct uptake as inorganic sulphate largely. Sulphate is a main inorganic form in most soils, although elemental and sulphide forms may present in soil under anaerobic conditions. Sulphate may be present in the soil solution, absorbed on soil surface or a insoluble compounds such as gypsum or associated with calcium carbonate. Out of above only the solution and adsorbed forms of sulphate are the primary pool of S in soil that are readily available to plants.

Principle

The methods available for accurate determination of total S in soils involve two steps:

(i) Conversion of the various S compounds in soils to one form, either by oxidation to sulphate (SO_4^{2-}). or by reduction to sulphide [Conversion to sulphate is more common than conversion to sulphide or by reduction to sulphides] and extraction with 1.0 percent sodium chloride. The sulphate produced is determined most commonly by turbidimetric procedure.

Apparatus

- Spectrophotometer or Calorimeter or Autoanalyzer
- Oven & Hot plate
- Volumetric flasks 50 mL and Beaker 100 mL
- Water both.

Reagents

- Magnesium nitrate solution: Dissolve 25.0 g magnesium metal in 400 ml of concentrated nitric acid in a 1,500 ml Erlenmeyer flask. If the metal does not dissolve completely add a further 50 ml of acid. Dilute with 100 ml of water to dissolve any crystals of magnesium nitrate, cool and dilute to 1500 ml.
- Nitric acid —24% v/v.
- Acetic acid—50% v/v.
- 0-phosphoric acid (concentrated).
- Barium chloride dihydrate ($BaCl_2$. $2H_2O$) : Crystals 20-30 mesh.
- Standard sulphate solution: Dissolve oven dried AR grade 0.5434g K_2SO_4, in 1000 ml of distilled water. This solution contains 100 ppm or 100 mg S lit^{-1}. Dilute 10 times to obtain 10 ppm solution.

Procedure

- Mix 1.0 g of soil with 2 ml of magnesium nitrate solution in a 100 ml beaker and evaporate to dryness at 70°C on an electric hot plate.
- Heat the residue overnight in a stainless steel oven at 300°C.
- Cool and add 5 ml of 25% nitric acid.

- Cover the beaker and digest on a water-bath for 2 hours and 30 min.; avoid loss of acid.
- After cooling dilute the contents of beaker with water and transfer to a 50 ml volumetric flask. Make up the volume.
- Take a 5 ml aliquot of the above solutions in a 50 ml volumetric flask and add 5 ml of acetic acid, 1 ml each of phosphoric acid and barium chloride.
- After 1 to 3 minutes measure the turbidity on colorimeter using blue filter or on spectrophotometer at 440 nm. The turbidity remain constant for 5-10 minutes.
- Preparation of standard curve : Transfer 0, 5, 10, 15 and 20 ml of 10 ppm stock solution into 150 ml Erlenmeyer flask and make up the volume with distilled water. Keep the flasks ready for making standard curve. Add 25 ml of stabilizing solution and 0.2 to 0.3g $BaCl_2 . 2H_2O$ crystals by a spatula and shake the flask for one minute. After 1-3 minutes measure the turbidity in a spectrophotometer. The turbidity remains constant during 3-10 minutes.
- Draw the standard curve and record the readings for samples.

Calculation

S content in soil sample (ppm) = S x Dilution Factor = S x 50/5 = 10S

Where, S = gS in aliquot obtained from standard curve

Total S in soil (%) = ppm S in soil x 10^{-4}

SO_4 = S x 3.0

1.19. Determination of Available Zinc, Iron, Manganese and Copper in Soil

Principle

In today's intensive agriculture, deficiencies of micronutrients, particularly, Zn,Cu, Fe and Mn have become one of the important factors limiting yields of crops in many parts our country. In soils having deficiency of any of these micronutrients, application of the deficient micronutrient causes a significant increase in crop yield. Chelating agents offer great promise for assessing readily available micronutrient cations in soils. These agents combine with free metal ions in solution to form soluble complexes DTPA offers the most favourable combination of stability constants for the simultaneous complexing

Zn, Cu, Mn and Fe. To avoid excessive dissolution of $CaCO_3$ which can release occluded micronutrients (not available for plants), the extractant is buffered in a slightly alkaline pH range and in part by including soluble Ca^{2+} triethanolamine (TEA) is used as buffer because it burns cleanly during atomization and has a pka = 7 .8. At the selected pH of 7.3, three fourth of TEA is protonated and is present as $HTEA^+$ When the extractant is added to the soil, addition Ca^{2+} and some Mg^{2+} enters the solution, largely because the protonated TEA exchanges with Ca^{2+} and Mg^{2+} from soil exchange sites. This raises the concentration of ionic Ca^{2+} by two to three fold and aids in suppressing the dissolution of $CaCO_3$ in calcareous soils with a 2:1 solution to soil ratio. The capacity of DTPA to complex each of the micronutrient cations is 10 times its atomic weight and ranges from 550 to 650 ppm depending on the micronutrient cation Thus, DTPA is present in excess of the micronutrient metal cations that are normally solubilized during an extraction. This excess reduces the possibility that the extraction of one micronutrient might significantly affect the amount of other micronutrient extracted.

Apparatus

- Atomic Absorption Spectrophotometer (AAS) .
- Mechanical Shaker.
- Volumetric flask etc.

Reagents

1. DTPA = 0.005 M (formula weight 393.35)
2. $CaCl_2 . 2H_2O$ = 0.01 M solution,
3. TEA = 0.1 M solution.

Extracting Solution

To prepare 1 litre of DTPA extracting solution, dissolve 13.1 ml reagent grade TEA, 1.967 g DTPA (AR grade) and 1.47 g of $CaCl_2. 2H_2O$ in 100 ml of glass distilled water. Allow some time for the DTPA to dissolve and dilute to approximately 900 ml. Adjust the pH to 7.3 ± 0.05 with 1:1 HC1 while stirring and dilute to 1 litre. Addition of approximately 4 ml of 1 N HCl will bring the pH of the solution to 7.3. This solution is stable for several months.

Standard Solutions

Zinc Standard Solution : Dissolve 0.439 g AR grade $ZnSO_4. 7H_2O$ in 200 ml of glass distilled water in beaker. Add 5 ml of 1:5 H_2SO_4. Transfer to a litre measuring flask and make volume to the mark to have a standard solution of 100 ug Zn ml^{-1} (100 ppm). Transfer 10 ml of this standard solution to 100 ml volumetric flask and dilute to the mark with DTPA extracting solution to have a stock solution of 10 ug Zn ml^{-1} (10 ppm). For preparing working standards, transfer, 1, 2, 4 and 5 ml of stock solution (10 ug Zn ml^{-1}) to a series of clean 100 ml volumetric flasks and dilute each to the mark with DTPA extracting solution. Volume of stock Zn solution taken: 0, 1, 2, 4, 6 ml Concentration of Zn now in solution:0,0.1, 0.2, 0.4, 0.6 ug ml^{-1} (ppm)

Iron Standard Solution: Dissolve 0.702 g of AR grade ammonium ferrous sulphate $(NH_4)_2$ SO_4 Fe_2SO_4 $6H_2O$ in 300 ml deionized or glass distilled water in a beaker. Add 5 ml of 1:5 H_2SO_4. Transfer to a litre measuring flask and make volume to the mark. This is a standard solution of 100ug Fe ml^{-1} (100 ppm). To prepare working standards, transfer 1, 2, 4 and 6 ml of stock solution and dilute each to the mark with DTPA extracting solution. Volume of stock Fe solution taken : 0, 1, 2, 4, 6 ml. Concentration of Fe in solution: 0, 1, 2, 4, 6 ug ml^{-1} (ppm)

Manganese Standard Solution: Dissolve 0.288 g potassium permanganate ($KMnO_4$) AR grade in 300 ml deionized water in a beaker. Add 20 ml concentrated H_2SO_4, warm at about 60°C and add oxalic acid solution drop-wise to make the solution colourless. Cool and transfer to a 1 litre measuring flask and make up the volume upto the mark. This solution contains 100 ug Mn ml^{-1} (100 ppm). To prepare working standards, transfer 1, 2, 4, 6 and 8 ml of the standard solution to a series of clean 100 ml volumetric flasks and dilute each to the mark with DTPA extracting solution. Volume of stock Mn solution taken : 0, 1, 2, 4, 8 ml Concentration of Mn in solution: 0, 1, 2, 4, 8 ug ml^{-1} (ppm)

Copper Standard Solution : Dissolve 0.392 copper sulphate ($CuSO_4$. $5H_2O$) of AR grade in 400 ml glass distilled water in a beaker. Transfer to a litre measuring flask and make up volume to the mark with glass distilled water. This is a standard solution containing 100 ug Cu ml^{-1}. To prepare the working standards, transfer 1, 2, 4 and 6 ml of stock solution to a series of clean 100 ml volumetric flasks and dilute each upto the mark with DTPA extracting solution.

Volume of stock Cu solution taken: 0, 1, 2, 4, 6 ml Concentration of Cu in solution: 0, 1, 2, 4, 6 4 ug ml^{-1} (ppm)

Standard curves and calibration: Usually at least 3 to 4 standards and a blank of each micronutrient cation are used for drawing a calibration curve. The blank solution (0 ug ml^{-1}) is used to zero the atomic absorption spectrophotometer. The standards are then analysed with lowest concentration first, and the blank run between standards to ensure that the base line (zero point) has not changed. A graph of absorbance versus concentration of standard solutions is plotted on a graph paper. The above calibration can also be performed in the concentration mode in which case the concentration of the sample is read off directly on the instrument. After setting the instrument, aspirate the sample, read the absorbance and find out the concentration against absorbance from the curve. Suppose it is M, than concentration of element in soil will be = M x dilution factor ug g^{-1} or ppm.

Procedure

(1) Weigh 10 g of finely ground air-dired soil in a 250 ml conical flask or polypropylene bottle. Then add 20 ml of the DTPA extracting solution.

(2) Cork the bottles or flask and place them upright on a horizontal shaker.

(3) Shake for two hours (120 minutes) with a speed of 120 cycles per minute.

(4) Filter the suspension through Whatman number 42 filter paper.

(5) Keep the filtrate in polypropylene bottles to be analysed for Zn, Cu, Mn and Fe with an Atomic Absorption Spectrophotometer (AAS).

(6) When samples need dilution before measurement, they should be diluted with DTPA solution to maintain a constant matrix.

Note

1. Experimental conditions such as shaking time, DTPA concentration, pH and temperature during shaking influence the amount of Zn, Cu, Mn and Fe extracted by DTPA (Table).
2. The most suitable pH of extracting solution is 7.3, shaking time 2 hours (120 minutes) and temperature during shaking 25 ± 1°C.
3. The values of the nutrient extracted will change if these precautions are not followed.

Calculations

Concentration of element in soil = M x 2 (Or Dilution factor; 20/10 = 2)ug g^{-1} or ppm Available element (kg/ha) = concentration of element (ppm) x 2.24

Discussion.

1. Critical limits of available Fe, Mn, Zn and Cu by DTPA extraction are 2.5—4.5, 2.0, 0.6 and 0.2mg kg^{-1} soil, respectively (Katyal and Ratan, 2003).

2. Critical levels of deficiency (mg kg^{-1} in plant tissue) in plants are : Zn 10-20, Fe 50, Mn 15-25 and Cu 2-5 (Katyal and Rattan, 2003).

3. Functions of reagents

(i) DTPA (Diethylene Triamine Penta Acetic Acid): Chelating agents offer great promise for assessing readily available micronutrient cations in soils. These agents combine with free metal ions in solution to form soluble complexes. DTPA offers the most favourable combination of stability constants for the simultaneous complexing of Zn, Cu, Mn and Fe. (This extractant had been adopted by most soiltesting laboratories in India).

(ii) Calcium chloride ($CaC1_2 . 2H_2O$): The excessive dissolution of $CaCO_3$ can release occluded micrornitrients (not available for plants). To avoid this the extractant is buffered in a slightly alkaline pH range and in part by including soluble Ca as $CaCl_2$.

(iii) TEA (Triethanolamine) It is used as a buffer because it burns cleanly during atomization and has a pKa = 7.8. At the selected pH of 7.3 about 75% of TEA is protonated and is present as HTEA. When the DTPA extractant is added to the soil, HTEA exchanges Ca^{+2} and Mg^{2+} from the soil and this raises the concentration of Ca^{2+} by two to three fold, which suppresses the dissolution of CaCO in calcareous soils.

4. The DTPA is present in excess of the micronutrient cations that are normally extracted.

5. The capacity of DTPA to complex each of the micronutrient cation is 10 times of its atomic weight and ranges from 550 to 650 ug g^{-1}.

Table: Effect of shaking time, temperature and pH on DTPA extractable micronutrients

Parameters	Nutrient (mg kg^{-1} soil)			
Zn	Zn	Cu	Mn	Fe
Shaking time (hours)				
1.0	0.26	0.33	7.6	3.4
3.0	0.28	0.41	10.1	4.4
16.0	0.47	0.68	20.0	7.7
Temperáture (°C)				
15	0.32	0.35	12.1	3.67
25	0.34	0.45	18.7	5.52
35	0.53	0.66	28.4	7.01
pH of extractant				
7.0	0.28	0.35	9.8	4.3
7.3	0.28	0.35	9.0	4.1
7.9	0.28	0.35	7.0	2.3

Source: Lindsay and Norvell (1978)

1.20. Determination of Available Boron in Soil

Principle

Boron in soil exits in organic an inorganic forms. The compounds of boron, which have high availability are water soluble. Boron deficiency is encountered in highly calcareaous and acid soils, while toxicity in salt-affected soils as well as soils irrigated with high B water. Soil extraction with cold or hot water continues to be the most acceptable method for determining the available B in soils. In soil extract, B can be determined by azomethine-H method (John et al. 1975). The method employs azomething-H the reagent to form a stable coloured complex of H_3BO_3 at pH 5.1 in aqueous media, which retains proportional absorbance — concentration properties for several hours independent of the presence of a wide variety of salts. The maximum absorbance occurs at 420 nm. This technique is rapid, reliable and more convenient to use than tradition procedures employing carmine, curcumin or quinalizarin.

Instruments

(i) Refrigerator

(ii) Colorimeter or spectrophotometer

(iii) Polypropylene test tubes of 10 mL or 15 ml. capacity

Reagents

i) Buffer masking solution: Dissolve 250 g of ammonium acetate and 15 g EDTA disodium salt in 400 ml distilled water and slowly add 125 ml of glacial acetic acid and mix.

ii) Acid solution. Fresh reagent should be prepared each week and stored in a refrigerator.

iii) Standard solution : To prepare the standard stock solution, dissolve 0.570 g boric (H_3BO_3) AR grade in 1 litre of distilled water to obtain a stock solution of 100 mg B/ml. Take 5 ml of this stock solution in a 100 ml volumetric flask and dilute to the mark. solution contains 5 mg B/ml.

Working Standards and Standard Curve

To a series of 25 ml volumetric flasks, add 0,0. 25,0.5, 1,2 and 4 ml of 5 mg B/ml solution. To each volumetric flask, add 2 ml. of buffer masking solution and mix, add 2 ml of Azomethine — H reagent solution. Allow to stand at room temperature for 30 minutes. Make the volume with distilled water and measure the absorbance at 420 mm on spectrophotometer. The concentration of B in working standard would be 0, 0.05, 0.10, 0.20, 0.40, 0.80 mg B/ml.

Over a concentration range of 0.5 to 10 ug B/ml, Azomethine — H solution forms a stable complex with H_3BO_3 at pH 5.1. Maximum absorbance occurs at 420 mm with little or no interference from a wide variety of salts. Plot a graph of concentration vs absorbance on a semilog paper, plotting concentration on X axis and absorbance on Y axis. Include distilled water for the 0.0 mg B/ml standard solution.

Extraction and determination of Boron

A 25 g soil sample, 50 ml of water and about 0. 5 g of actwated charcoal is boiled for 5 minutes in quartz flask and filtered immediately through Whatman filter paper no. 42. Take 5 ml of the extract in a 25 ml volumetric flask and 4 ml of buffer masking solution and 4 ml of azomethine — H reagent solution is added. The colour is allowed to develop for 1 hour, and the volume is made to the mark. Intensity of colour is measured spectrophotometrically at 420 nm.

Calculation

Boron (ug/g or ppm) = Concentration of B in analyzed sample in soil sample (reading from the standard curve) x 10.

1.21. Determination of Available Molybdenum in Soil

Molybdenum is required in the smallest quantity among the essential nutrients and absorbed in form of MoO_4^- ion. Its availability in high is the alkaline pH range,

Principle

Ammonium oxalate (pH 3·3) or Grigg's reagent is considered to be the best one for determination of available Mo. Further, this extractant is easy to prepare, has sufficient buffering capacity to prevent any material change in the pH of the soil extract and forms stable complexes with Mo [$MoO_3 . C_2H_2O_4$ and $(MoO_3)_2 . C_2H_2O_4$]. The molybdate absorbed on soil colloids and clay is presumably replaced by the oxalate ions. This exchange is made irreversible by the formation of strong Mo-oxalic acid complexes to make a single extraction effective.

Apparatus

1. Spectrophotometer or colorimeter
2. Hot plate

3. Refrigerator
4. Water bath.

Reagents

1. Potassium iodide solution (50%) : Dissolve 50.0 g of potassium iodide in 100 ml of double distilled water (DDW).
2. Ascorbic acid solution (50%) : Dissolve 50.0 g ascorbic acid in 100 ml of DDW.
3. Sodium hydroxide solution (10%): Dissolve 10.0 g of NaOH in 100 ml of DDW.
4. Thiourea solution (10%) : Dissolve 10.0 g in 100 ml of DDW and filter. Prepare fresh before use.
5. Toluene-3, 4-dithiol solution (commony called dithiol) : Weigh 1.0 g of AR grade melted dithiol (51°C) in a 250 ml glass beaker add 100 ml of the 10% NaOH solution and .warm the content upto 51°C with frequent stirring for 15 minutes. Store in a refrigerator.
6. 'Tartaric acid solution (10%): Dissolve 10.0 g tartaric acid in 100 ml of DDW.
7. Iso-amyl acetate solution.
8. Ethyl alcohol for washing of glasswares.
9. Ferrous ammonium sulphate solution: Dissolve 64.0 g of the ferrous ammonium sulphate in about 500 ml of DDW and then make up the volume to one litre.
9. Extracting reagent : Dissolve 24.9 g of AR grade ammonium oxalaté and 12.6 g oxalic acid in water and make up the volume to 1 litre. Adjust the pH at 3.3 either with diluted HCl or diluted ammonia solution.
10. Standard stock solution (100 mg Mo L^{-1}) : Dissolve 0.150 g of AR grade MoO_3 in 100 ml of 1.1 N NaOH make slightly acidic with diluted HCl and make up the volume to 1 litre.
11. Working standard solution (1 mg Mo L^{-1}): Dilute 10 ml of the stock solution to litre procedure

Soil & Extraction

1. Weigh 25 of air-dried soil sample in a 500 ml corning or pyrex glass. Add 250 ml of the extracting solution (1: 10 ratio) and shake for 10 hours over an end to end horizontal shaker.
2. Filter through Whatman number 50 filter paper. Collect 200 ml of the clear filtrate in a 250 ml glass beaker and evaporate to dryness on a water bath.
3. Heat the contents at 500°C in a muffle furnace for 5 hours to destroy organic matter and oxalates. Keep it overnight.
4. Digest the contents with 5 ml of HNO_3—$HClO_4$ mixture (4:1), then with 10 ml of 4N H_2SO_4 and H_2O_2 each time bringing to dryness.
5. Add 10 ml of 0.1 N HCl and filter. Wash the filter paper with separate 10 ml of 1 N HCl and DDW until the volume of the filtrate is 100 ml.
6. Run a blank side by side (without soil)

Colour development

1. Take 50 ml of the filtrate in 250 ml separatory funnels and add 0.25 ml of ferrous ammonium sulphate solution and 20 ml of DDW and shake vigorously.
2. Add excess of KI solution and clear the liberated iodine by adding ascorbic acid drop by drop while shaking vigorously.
3. Add 1 ml of tartaric acid and 2 ml of thiourea solution and shake vigorously.
4. Add 5 drops of dithiol solution and allow the mixture to stand for 30 minutes.
5. Add 5 ml of iso-amyl acetate and separate out the contents (green colour) in colorimeter tubes/cuvettes.
6. Read the colour intensity at 680 nm (red filter).

Preparation of standard curve for Mo

1. Measure 0, 2, 5, 10, 15 and 20 ml of the working standard Mo solution containing 1 mg Mo l^{-1} in series of 250 ml separatory funnels.
2. Proceed for colour development as described above for sample aliquots.
3. Read the colour intensity at 680 nm (red filter) on colorimeter and prepare the standard curve by plotting Mo concentrations against readings.

Calculation

Available Mo in soil (ppm or ug g^{-1}) $\frac{R \times 250 \times 1}{200 \times 25} = \frac{R}{20}$

R = Mo concentration in ug g^{-1} or ppm obtained by from standard curve against sample reading

1.22. Analysis of Gypsum Requirement

Principle

Soils with pH above 8.5 have abundance of exchangeable Na^+, which needs to be exchanged with Ca^{++} (from gypsum, $CaSO_4.2H_2O$) or H^+ (to be obtained from elemental S or pyrites FeS_2 after their oxidation to H_2SO_4. Gypsum requirement of sodic/alkali soils is determined by treating the soil with a known amount of excess saturated gypsum and solution and then estimating the unreacted or unutilized by titration with EDTA solution (Schoonover, 1952).

$$Na^+ - clay + CaSO_4 \rightarrow clay + Na_2SO_4 \downarrow$$

leached

Apparatus

1. Mechanical shaker.
2. Erlenmeyer flask
3. Burette

Reagents

1. Saturated gypsum solution : Dissolve 5.0 g $CaSO_4$. $2H_2O$ in 1 litre of distilled water.
2. $CaCl_2$ (0.01 N) : Dissolve 0.5 g AR grade $CaCO_3$ powder in about 10 ml of diluted (1:3) HCl and make up the volume to 1 litre. This is the standard solution.
3. EDTA solution (0.01 N) : Dissolve 2.0 g of EDTA-disodium salt and 0.05 g AR grade $MgC1_2$ in about 50 ml water and make up the volume to 1 litre. Titrate an aliquot of this reagent with 0.01 N $CaCl_2$ solution to standardize it.
4. Eriochrome black T (EBT) indicator : Dissolve 0.5 g EBT dye and 4.5 g hydroxylamine hydrochloride in 100 ml of 95% ethanol.

5. NH_4OH—NH_4C1 buffer : Dissolve 67.5 g NH_4CI in 570 ml concentrated ammonia solution and dilute to 1 litre. Adjust the pH at 10 using diluted HCI/diluted NH_4OH.

Procedure (Singh et al, 1999)

1. Transfer 5.0 g soil in a 250 ml Erlenmeyer (conical) flask.
2. Add 100 ml saturated gypsum solution and stopper to the Erlenmeyer flask.
3. Shake the contents of the Erlenmeyer flask for 5 minutes on a mechanical shaker.
4. Filter the contents of the Erlenmeyer flask through Whatman No. 1 filter paper.
5. Transfer 5 ml filtrate into a 100/150 ml porcelain dish.
6. Add 1 ml of the NH_4OH - NaCl buffer solution and 2 to 3 drops of EBT indicator.
7. Titrate the solution in the porcelain dish with a thin 0.01 N EDTA solution.
8. Run a blank titration with 5 ml of saturated gypsum solution.

Calculations

Ca or Ca + Mg (me l^{-1}) in the aliquot = 2V

Where V stands for volume of EDTA solution used (sample reading—blank reading)

Since l litre extract = 50 g soil (5 g soil to l00 ml)

Ca requirement (me 100 g^{-1} soil) or G

G = [2V for added gypsum solution —2 V for ifitrate] x 2

Gypsum requirement of soil in tons/ha upto 30 cm soil depth = G x 3·852

Note

1. Apply correct amount depending upon the purity of gypsum.
2. Depending upon the percentage of exchangeable Na^+, soils are classified in relation to sodicity hazard as given in Table.

Some of these are given below:

(i) Tolerant: barley, sugarbeet, rice (due to water submergence), cotton, lucerne.

(ii) Moderately tolerant: Sorghum, soybean.

(iii) Moderately sensitive: Sugarcane, corn, peanut, potato, radish, spinach, sweet potato, tomato, turnip.

(iv) Sensitive: Pulses, onion, carrot.

Exchangeable Sodium percentage (ESP) and sodicity hazard

ESP	Sodicity hazard
<15	None to slight
15-30	Light to moderate
30-50	Moderate to high
50-70	High to very high
>70	Extremely high

Source: Abrol *et al.* (1988)

2

Plant Analysis

Accumulation of nutrient elements in plant tissues indicates the accessibility of the concerned elements from the soil to the plant. Although different plant species and even different varieties of the same species may vary in their nutrient requirements, composition of a part or of the whole plant may well be adopted as a guideline to support the soil-test results for factual assessment of soil's nutrient supplying power. Also, the concentration of certain heavy metals in plant might be a result of their solubility in soil and may lead to polluting effects. To a certain extent, there exists a good relationship between the concentration of an element in the plant and the total biomass of the plant. A correct balance of nutrients in the plant tissues is closely *associated with the maximum yield,* except in the cases of luxury consumption of nutrients like potassium.

Plant analysis as an aid to soil testing has its limitations too. These, among other things, include the risk of sampling error as well as the chances of wrong interpretation of the test results for making recommendation. Significant error in the plant analysis may arise due to wrong sampling alone, especially when the plant analysis aims at diagnosis of nutrient deficiency in standing crops. Selecting the right plant part, stage of growth and time of sampling are very crucial in plant analysis.

2.1 Plant Sampling

For a meaningful plant analysis, utmost care should be exercised in plant sampling. Whole plant analysis is done for working out the total nutrient uptake, which is usually carried out on the aboveground (shoot) material. Analysis of roots may either be taken up separately or individual plant parts like leaf, petiole *etc.* are sampled for the purpose of diagnosing nutrient deficiency in perennial fruit trees. Thus, depending on the purpose of analysis, plant sampling needs to be planned. Table 2.1 shows the specific tissues to be sampled from different plant species.

Table 2.1: Plant parts recommended for sampling

Crop	Part to be sampled with stage/age
Grain, pulses, oilseeds, fibre and commercial crops	
Wheat	Rag-leaf, before head emergence
Rice	3 rd leaf from apex, at tillering
Maize	Ear-leaf, before tasseling
Barley	Flag-leaf at head emergence
Oat	Flag-leaf, before inflorescence emergence
Pulses	Recently matured leaf, at bloom initiation
Groundnut	Recently matured leaflets, at maximum tillering
Sunflower	Youngest mature leaf blade, at initiation of flowering
Mustard	Recently matured leaf, at bloom initiation
Soybean	3rdleaf from top, after 2 months of planting
Cotton	Petiole, 4th leaf from the apex, at initiation of flowering
Jute	Recently matured leaf, at 60 days age
Sugarcane	3rd leaf from top, after 3-5 months of planting
Sugarbeet	Petiole of youngest mature leaf, at 50-80 days age
Tea	leaf from up of young shoots
Coffee	3rdor 4thpair of leaf from apex of lateral shoots, at bloom
Vegetables	
Potato	Most recent, fully developed leaf (half-grown)
Tomato	Leaves adjacent to inflorescence (mid-bloom)
Onion	Top non-white portion (⅓to ½ grown)
Brinjal	Blade of most recent, fully developed leaves
Beans	Uppermost, fully developed leaves
Cauliflower	Most recent, fully matured leaf, at heading
Cabbage	Wrapper leaf at 2-3 month age
Pea	Leaflets from most recent, fully developed leaves, at first bloom
Carrot	Most recent, fully matured leaf, at mid-growth
Radish	Most recent, fully developed leaf
Turnip & Sugar beet	Most recent, fully developed leaf, at 30-50 days age
Spinach	5th leaf from tip (omit unfurled) at the stage of bud starting to small fruits
Cucumber	Most recent, fully developed leaf
Ornamental plants	
Bougainvillea, Jasmine, Croton, Fern, Ficus, Geranium, Gerbera, Gladiolus, Lilly, Orchid, Hibiscus & Rose	Most recent, fully developed leaves at the stage of flower bud of pea size.
Fruit crops	
Almond	3rd leaf from top, at the beginning of bloom
Apple, Pear	Leaves from middle of terminal shoot growth, 8-12 weeks after full bloom, 2-4 weeks after formation of terminal buds in bearing trees
Blackberry	Latest matured leaf from non-tipped canes, 4-6 weeks after peak bloom

Cherry	Fully expanded leaves, mid shoot current growth in July-August
Peach	Mid-shoot leaves, fruiting or non-fruiting spurs, mid-summer leaves, fruiting
Strawberry	Youngest, fully expanded matured leaf without petiole, at peak or harvest period
Plum	Leaves from middle of current season's extension growth, in January-February
Banana	Petiole of 3 open leaf from apex, 4 months after planting
Cashew	4th leaf from tip of matured branches, at beginning of flowering
Grapes	5th petiole from base at bud differentiation for yield, and petiole opposite to bloom for quality
Citrus	fruits 3-5 month old leaves from new flush, l' leaf of the shoot, in June
Guava	31 pair of recently matured leaves, at bloom (August or December)
Mango	Leaf with petiole (4-7 month old) from middle of shoot
Papaya	6th petiole from apex, six months after planting
Pineapple	Middle one-third portion of white basal portion of *4th* leaf from apex, at 4-6 months age
Pomegranate	8thleaf from apex at bud differentiation, in April and August
Falsa	4thleaf from apex, one month after prunnng
Ber	6th leaf from apex from secondary or tertiary shoot, 2 months after pruning.

Sources: Kensworthv (1964), Reuter and Robinson (1986), Jones *et al.* (1991), Bhargava and Raghupatbi (1993)

Procedure for Plant Sampling

1. To correct the nutrient problem, the plant tissue sampling procedure is given in Table-2.1.
2. To estimate nutrients uptake in research project, where each plot receiving different treatment as a separate unit it has to be sampled separately by picking up few representative plants at random from each plot.
3. Remove the shoot (aerial part) with the help of sharp knife or blade or scissors for whole shoot analysis or the desired plant part (as given in Table) for analysis of specific plant parts.
4. If roots are to be included uproot the whole plant carefully from wet soil, retaining even the fine/active roots. Gently dip plant roots in running water to remove adhering soil as far as possible.
5. Wash the plant part with water several times.
6. Wash the sample with 0·2% liquid detergent to remove waxy/greasy coating on the leaf surface, which is often present.

7. Wash with 0.1 N HC1, followed by thorough washing with water and give final rinse with distilled water.
8. Rinse with double distilled water specifically if micronutrient analysis is to be carried out.
9. Soak with dry tissue paper.
10. Air dry the samples on clean surface at room temperature for at least 2-3 days in dust free atmosphere.
11. Put the samples in oven and dry at 60°-70 for 40 hr or as given in following Table......

Plant Material	lime (Hours) for drying at	
	at 50°C	at 100°C
Milled cereals and compound feed	—	24
Cereals, grains	—	40
Oilseeds and mustard	—	18
Herbage, silage, hay	—	18
Potatoes	24	18
Bulkey roots, carrot	48	—

12. Grind the samples in stainless steel mill using 0.5 to 1.0 mm sieve. Clean the cup and blades of the grinding mill before each sample.
13. Put back the samples in oven and dry again for few hours more for constant weight.
14. Store in well stoppered plastic or glass bottles or paper bags for analysis.
15. Write the name, number and other information regarding samples on the storage bags or bottles may also put a tag containing this information into the bags or bottles.

Precautions

(1) Do not put fresh plant sample in polythene bags longer.

(2) Sufficient hole should be made on the bags having fresh sample for proper air circulation in bag.

(3) Do not sample the following:

(i) Young, emerging leaves; old, mature leaves and seed, these plant parts usually are not suitable because they are not likely to reflect in nutrient status of the whole plant.

(ii) Diseased or dead plant.

(iii) Plants' that have insect or mechanical damage.

(4) Brushes scissors and trays made of brass or copper must not be used in preparation of sample to be analyzed for trace elements.

(5) Clean out the mill and other equipment's, between samples in order to avoid cross contamination.

(6) Take the representative sample from homogeneous powders of the dried sample

2.2 Ashing or Digestion of Plant Sample for Elemental Analysis

For the release of mineral elements from plant tissues, both dry ashing and wet oxidation procedures are ridely practiced. The choice of ashing procedure depends mainly upon the availability of equipments, reagents and personal consideration of worker. Some merits and demerits of both procedures are given below:

Wet ashing/oxidation	Dry ashing
1. Boron completely and halides to great extent are lost.	1. Unless bases such as Mg NO_3 or Mg acetate are employed P and S are lost.
2. Used for determination on of P, K, Ca, Mg, S. Fe, Mn, Zn and Cu and not used for B&Mo	2. Used for determination of Na, K, Ca, Mg, Cu, Fe, Mn, Zn, B and Mo cannot be use For N, P and S.
3. It is time taking, requires more reagents and more attention of analyst. (Particularly requiring minimal attention of analyst. for disposal of acid fumes).	3. It is précised, easy and rapid method, requiring minimal attention of analyst.
4. May create pollution.	4. Pollution free.
5. It require more reagents, equipments and glass wares,	5. Plant material is directly ashed and requires only porcelain or silica dishes and muffle furnace.

Source : Prasad et al. 2006.

2.1.2 Wet digestion (Wet ashing/oxidation) Method

Wet digestion of plant sample can be done by di-acid mixture ($HC1O_4 + HNO_3$ 4:9) or a tri-acid mixture ($HNO_3 + H_2SO_4 + HC1O_4$: 9:4:1).

The tri-acid digestion is recommended when only P or K are to be estimated. Sulphur cannot be estimated from tri-acid extract due to presence of H_2SO_4. Similarly, Ca and Mg will also be under estimated.

Principle

The organic matter of Plant material is destroyed by wet digestion with a mixture of Nitric acid and Perchloric acids. The acids are removed by volatilization and any silica present is dehydrated and made in soluble. The soluble constituents are dissolved in H_2SO_4.

2.1.3 Di-acid Digestion

Apparatus: Hot Plate, Beakers of borosilicate glass etc.

Reagents

1. Conc. HNO_3 (AR grade)
2. 60 per cent $HC1O_4$ (AR grade)
3. Approx. 2N HCl (AR grade)

Procedure

1. Weigh 0.5 or 1.0 g of dried and processed plant sample in a 250 mL conical flask.
2. Add 10 mL of conc. HNO_3, place a funnel on the flask and keep for about. 6-8 hour overnight at a covered place/chamber for predigestion.
3. After pre-digestion when sample is no more visible, add 10 ml of conc. HNO_3 and 2-3 mL of $HClO_4$.
4. Keep on a hot plate in acid-proof digestion chamber having fume exhaust system and heat at about 100°C for first one hour and then raise the temperature to about 200°C.
5. Continue digestion until the contents become colorless and only white dense fumes appear.
6. Reduce the acid contents to about 2-3 mL by continuing heating at the same temperature. Do not allow to dry up.
7. Remove from hot plate, cool and add 10 mL of dilute and perfectly colourless HC1.
8. Warm slightly and filter through Whatman No. 42 filter paper into 100 mL Volumetric flask.
9. Give 3-4 washing of 15-20 mL portion of distilled water and make the volume to 100 ml.

2.1.4 Tri-acid Digestion

Apparatus

(i) Water bath

(ii) Hot plate

(iii) Beakers of borosilicate glass.

Reagents

Tri-acid mixture: Mix AR grade conc. HNO_3, H_2SO_4 and $HClO_4$ in 10: 1: 4 ratio and cool.

Procedure

1. Transfer 0.5 or 1.0 g of dried and processed plant sample to a 250 mL conical flask. Add 5 mL of conc. HNO_3.
2. Keep a glass funnel on the flask, place it on a water bath and heat at 100°C for about 30 minutes.
3. Shift the flask to a hot plate and heat at 180-200°C. Measure temperature in a flask containing glycerol kept on the hot plate.
4. Continue boiling until near to dryness but not drying completely. Cool and add 5 mL of the tri-acid mixture.
5. Heat at 180-200°C until the dense white fumes are evolved. Continue digestion until the mixture is largely volatilized.
6. If the contents are still brown, cool a little and add 3-4 mL of the tri-acid mixture and continue the digestion as described above. This is rare as there is still sufficient $HClO_4$ to oxidize the charred material.
7. Remove the flasks when only moist, clear and white contents are left. The entire quantity of $HClO_4$ has volatilized by this stage.
8. Cool and add about 50 mL of distilled water or double distilled water if micronutrients are to be analyzed.
9. Filter into 100 mL flask, giving washings to make the volume to 100 mL.
10. Use filtrate for analysis.

Note

(i) This procedure should not be used for sample type other than plant material

(ii) This procedure preferably be carried out in fume cup board.

2.1.5 Dry ashing

Principle

The organic matter of plant material is destroyed by dry combustion and the soluble mineral constituents in the ash are dissolved in hydrochloric acid. Any silica present in dehydrate makes insoluble.

Apparatus

- Muffle furnace capable of achieving 550°C ± 15°C
- Evaporating basins —30 ml, translucent silica, shallow form, with round bottom and spout crucible, hot plate, volumetric flasks.

Reagents

- Hydrochloric acid, approx. 36% m/m HC1.
- Hydrochloric acid, approx. 6 M — Mix equal volumes of hydrochloric acid, approx. 36% m/m HCI and water.

Procedure

1. Transfer 2 g of dried sample into a basin and place the basin in cool muffle furnace to 500°C and maintain this temperature (usually overnight) until a whitish — gray ash remains.
2. If difficulty is experienced in carbon free ash, remove the basin from muffle furnace, moisture the cold ash with water, dry thoroughly at 102°C and reheat at 500°C.
3. When all the organic matter has been destroyed remove the basin from the muffle furnace, cool and cover with a watch glass.
4. Add 10 ml of approx. 6 M hydrochloric acid, taking care that losses due to effervescence do not occur.
5. Remove and rinse the watch glass, collecting the washings in the basin.
6. Place the basin on a water bath and evaporate the solution to dryness.

7. When drying continue heating for 1 hour either on a water bath or in an oven at 102°C.
8. Moisten the residue with 2 ml hydrochloric acid, approx. 36% m/m HC1 — cover the basin with a watch glass and gently boil for 2 min.
9. Add approx. 10 ml of water and again boil.
10. Remove and rinse the watch glass, collecting the washings in the basin.
11. Quantitatively transfer the contents of the basin in to a 100 ml volumetric flask and dilute to 100 ml.
12. Filter through a Whatman 41 filter paper, reject the first few ml of filtrate and retain the remainder.
13. Carry out a blank determination.

Note

(1) The sample solution can be used for determination of elements immediately as prescribed, otherwise sample solution should be stored in polythene bottle.

(2) If trace elements are to be determined, it is necessary to make final volume of solution up to 10-20 ml only or increase plant tissue sample up to 2 g.

(3) If Boron is to be determined borosilicate free glass ware and laboratory ware must be used.

(4) Care should be taken when placing the sample containing HNO_3 in the muffle furnace

2.3. Determination of Total Nitrogen in Plant

Total N in plant samples in determined commonly by in Kjeldahl Method and by Colorimetric procedure

a) Kjeldhal Method (Modified to include Nitrates)

Principle

The plant sample is digested with sulfuric acid salicylic acid (30: 1) at a temperature between 360°-410°C, the digestion process is slow or incomplete but above 410°C some of ammonia may be lost. Organic and, nitrate nitrogen is converted to ammonium sulphate and ammonia gas is distilled into boric acid and titrated with standard sulfuric or hydrochloric acid. Nitrates present in sample forms nitrocompound by reaction of salicylic acid in acid medium. These nitro compounds reduced to corresponding amino compounds by heating mixture with sodium thiosuiphate. The rate of digestion is accelerated by using copper sulphate as a catalyst and anhydrous sodium sulphate or Potassium sulphate to raise the boiling temperature of sulfuric acid. To avoid the interference of water in nitration of salicylic acid, plant sample should be oven dried.

Apparatus

- Block digester 20 or 40
- Micro Kjeldhal apparatus
- Kjeldahl tubs 100ml
- Erlenmeyer flask 125 ml
- Digital or simple burette
- or Automatic digestion distillation cum titration or distillation system (programmable or ordinary).

A. Digestion

Reagents

1. Sulphuric salicylic acid: dissolve one g salicylic acid 30ml of Conc. H_2SO_4.
2. Sodium thio-sulphate: 24 mesh dried powder.
3. Sulphate mixture : 20 parts of K_2SO_4 + 1 part catalyst mixture (20 part $CuSO_4$ + 1 Part selenium powder).

Procedure

1. Put 0.2-0.5 g dried ground sample in 100 ml Kjeldhal tube.
2. Add 20 ml of sulphuric-salicylic acid mixture in the tube and swirl gently so as to dry sample come to the contact with mixture. It is allowed to stand overnight.
3. Add 5 g sodium thiosulphate on next day and heat gently for 5 minutes avoiding frothing.
4. After cooling the content, add 10 g of sulphate mixture and put the tubes in block digester at 150°C for 45 minutes and then increase up to 410°C by increasing temperature in split of 50°C at an interval of 30 minutes.

5. Bumping during the digestion can be avoided by adding of some glass beads.
6. When the sample is clear (approx. after 2-3 hr), cool it and then add few ml of water to dilute it.
7. The distillation is carried out as follows.

B. Distillation

Reagents

1. Boric acid (4 per cent): Dissolve 40 g H_3BO_3 in about 700 ml water. Add 20 ml mixed indicator. Adjust the pH to 5.0 with N/10 NaOH or 0.1 N HCl. Make up the volume to 1 liter. The concentration of this reagent need to be precise as long as the amount of boric acid is more than the chemically equivalent to the amount of ammonia to be absorbed.
2. Mixed indicator: Dissolve 0.3 g of bromocresol green and 0.2 g methyl red in 400 ml of 90% ethanol. The indicator colour will change from red in acid solution to blue in alkaline solution. 3. Sodium hydroxide (40 per cent): Dissolve 400 g of technical grade NaOH in a beaker containing 600 ml of distilled water. Place the beaker in a cold water bath to dissipate the heat produced. When cool, dilute to 1 liter and store the solution in a screw-top bottle.
4. Sodium carbonate Transfer 10 to 20 g of AR grade Na_2CO_3 to a Pyrex beaker and heat at 270° for 3 hrs. Cool the beaker in a desiccator.
5. Methyl Orange indicator: Dissolve 0.1g of methyl orange in 100 ml distilled water.
6. Standard hydrochloric acid (0.1 N): Dilute 9 ml of concentrated HC1 to 1 litre with distilled water. Standardize this approximate 0.1 N HC1 solution as follows : Dissolve exactly 0.530 g of sodium carbonate reagent in 20 ml distilled water. Dilute to 100 ml. Transfer 10 ml of this 0.1 N sodium carbonate solution to 125 ml Erlenmeyer flask. Add two drops of methyl orange indicator. Titrate the approximate 0.1 N HCl solution into the 0.1 N sodium carbonate until the methyl orange indicator turns reddish orange. Boil the solution gently for one minute and then cool to room temperature by running tap water over the outside of the flask. If the colour changes back to orange, titrate more HCl until the first faint but permanent reddish orange colour appears in the solution.

$$\text{Normality of HCl} = \frac{0.1\text{x}10}{\text{ml of HCl used in titration}}$$

7. Standard hydrochloric acid (0.005 N) : Transfer 50 ml of the standardize 0.1 N HC1 to a 1.0 litre volumetric flask and make up to volume with distilled water.

Procedure

Distillation

1. Transfer the digested sample from Kjeldahl flask into the Kjeldahl micro distillation apparatus. Rinse the flask thrice with distilled water, each time emptying the rinse water into the distillation apparatus. Use a minimum amount of water.
2. Prepare a 125 ml Erlenmeyer flask containing 10 ml of 4 percent boric acid reagent. Place the flask under the condenser of the distillation apparatus and make sure that the tip of the condenser outlet is beneath the surface of the solution in the flask.
3. Allow steam from the boiler to pass through the sample and add slowly approx. 10 ml of 40 percent NaOH to the distillation apparatus. This will lead distilling off the ammonia into the flask containing boric acid.
4. Distillate the sample for 7-10 minutes. Then lower the flask and allow the solution to drop from the condenser for about 1 minutes. Wash the tip of the condenser outlet with distilled water.

C. Titration

Titrate the solution of boric acid and mixed indicator containing the distilled off ammonia with the standardized HC1.

Note:

(a) Use the standardized 0.1 N HCl for samples containing 1.5 to 4 per cent nitrogen.

(b) Use the standardized 0.05 N HCl of samples containing less than 1.5 per cent nitrogen.

(c) Try to have a titration value of more than 2 ml so that the titration error will be negligible.

(d) Determine the titration value of a blank solution of boric acid.

Calculation

$$\text{Nitrogen \% m sample} = \frac{(ST - BT) \text{ x normality of standard acid X 14 X 100}}{\text{Sample Weight (g) x 1000}}$$

$$= \frac{(ST - BT) \text{ x normality of HCl (0.1) x 1.4}}{\text{sample Weight (g)}}$$

$$= \frac{(ST - BT) \text{ x 0.14}}{\text{sample weight (g)}}$$

Where

BT = Titration value of Blank

ST = Titration value of sample

b)*Colorometric Method*

Nitrogen in plant sample can also be estimated colorometrically either manually or by using Auto Analyser System (Bacthgen and Alley 1989; Linder, 1942)

Apparatus

1. Spectrophotometer
2. Block Digester 20-40
3. Kjeldahl tubes 100 ml
4. Volumetric flasks

(a-1)Colorimetric Method of Bacthgen and Alley

Reagents

1. **Sodium salicylate sodium nitroprusside solution:** 150 g of Na-salicylate and 0.03 g of Na-nitroprusside are dissolved in 1 Litre of distilled water. Store in dark bottle.
2. **Sodium hypochlorite solution** : Take 6 mL of 5.25% Na-hypochlorite solution in 100 mL volumetric flask and dilute it to the volume with distilled water. Prepare this solution fresh every day.
3. **Working buffer solution** : (0. 1 M $Na_2HPO_4$5% Na-K tartarate, 5.4% NaOH) :26.8 g Na_2HPO_4. $7H_2O$ is dissolved in 600 mL of deionized or distilled water. Add50 g Na-K tartarat and 107 g 50% solution of NaOH. Dilute to 1 Litre with distilled water.

4. **Standard solution** : Dissolve 0.1185 g of dry $(NH_4)_2 SO_4$ in distilled water and make volume to 1 Litre. This will give standard solution of 250 mg N/Litre. Pipette out 0, 0.5, 1, 1.5, 2, 3, 4 and 5 mL of this solution in 25 mL volumetric flask and make volume to prepare standard solution containing 0, 5, 10, 15, 20, 30, 40 and 50 mg N/Litre (ppm).

Procedure

(A) Digestion

The digestion of sample is earned out m the same manner as stated in modified Kjeldahl method except salt catalyst mixture This mixture is prepared by mixing 30 parts of K_2SO_4+ 1part of HgO as catalyst and approximately 1 g of it is added to each digestion tube or flask.

(B) Determination of N

1. Take 1 mL of aliquot from plant digest m 25 mL Volumatic flask.
2. Add 5.5 mL of working buffer solution and stirred thoroughly.
3. Then add 4 mL of Na-salicylate Na-nitroprusside solution and mixed
4. Now add 1 mL of Na-hypochlorite solution and mixed again thoroughly. The reagents should be added in the above order otherwise solid residues may be formed.
5. The solution is allowed to stand for 45 minutes to ensure complete colour development. The absorbance is then read on a spectrophotometer at 650 nm.
6. Preparation of standard curve: To each standard solution colour is developed by procedure given above. Draw a graph of mg N/L (ppm N) verses absorbance.

Calculation

Weight of plant sample =0.5g

Volume of digestion = 100 mL

Aliquot taken = 1 mL

Concentration of N obtained from standard curve = C

Per cent N m plant sample = $\frac{\text{Cx } 100 \text{ x } 25}{0.5 \text{ x } 1 \text{ x } 10000}$

C is concentration of N in mgfL obtained from standard curve.

(a-2) Colorimetic Method of Linder

Reagents

1. Concentrated H_2SO_4 AR Grade
2. H_2O_2 (30%)
3. NaOH (10% aqueous solution)
4. Sodium silicate (10% aqueous solution)
5. Nessler's reagent
6. Standard ammoniurn sulphate solution: dissolve 0.1185 g of $(NH_4)_2SO_4$ in distilled or deionized water and make volume to 1 L.

Procedure

(A) Digestion

Digest 0.1 g plant material by H_2SO_4 and H_2O_2 (wet digestion) and make final volume of digest to 100 mL

(B) Determination of N

1. hike 5 mL of aliquot (plant digest) in 50 mL volumetric flask.
2. Add 2 mL and 1 mL of 10% solutions of sodium hydroxide and sodium silicate, respectively.
3. Add distilled or deionized water upto 45 mL and then 2 mL of Nessler's reagent.
4. Make the final volume (50 mL) with distilled or deionized water.
5. Run a blank following the same procedure.
6. Adjust colorimeter using blank and take reading at 540 nm or blue filter.

(C) Preparation of Standard Curve

Pipette out 0, 1, 2, 3, 4, 6, 8 and 10 mL of standard ammonium sulphate solution in 50 mL volumetric flask. Colour is developed by repeating steps 2-4 given above. Take reading on colorimeter. Draw a graph between N (mg/L) vs absorbance.

Calculation

Weight of plant sample = 0. 1 g

Volume of plant digest = 100 mL

Volume of aliquot taken = 5 mL

C = Concentration of N in aliquot as read out from the curve against the corresponding reading (mg/L)

d.f = dilution factor

C x d.f C x Cx 100 x 50

$$\text{Percent N} = \frac{C \times d.f}{10000} \text{ or } \frac{C \times C \times 100 \times 50}{5 \times 0.1 \times 10000}$$

2.4. Determination of Phosphorus in Plant

Colorimetric Estimation: Plant sample can be digest in di-acid (HNO_3 — $HClO_4$) or tri-acid (HNO_3 — H_2SO_4 — $HClO_4$) mixture.

Principle

Plant phosphorus is converted into orthophosphates during digestion. This orthophosphate reacts with vanadate and molybdate and gives yellow coloured vanadomolybdo-phosphoric hetropoly complex in acid medium. The yellow colour is attributed to a substitution of oxyvanadium and oxymolybdnum radicals for the oxygen of phosphates. The intensity of colour is directly proportional to concentration of phosphates presents in the sample which can be read on spectrophotometer or colorimeter. The colour is developed in about 25-30 minutes and it is stable for 2 to 8 weeks.

Apparatus

1. Spectrophotometer or colorimeter.
2. Volumetric flasks

Reagents

- Ammonium metavanadate solution (0.25%): Dissolve 1.25g of ammonium metavanadate in 250 ml of boiling water, add 10 ml of concentrated HNO_3, dilute it to 500 ml and store in an amber glass bottle. The vanadate concentration is moderately critical (within 10%).
- Ammonium molybdate solution (5%) dissolve 25g of ammonium molybdate in 400ml distilled water (warmed to 50 °C) cool the solution, make up the volume to 500ml and filter if cloudy, store and amber glass bottle the molybdate concentration is not critical.

- Nitric acid (HNO_3) solution (5 N): Dilute 530 ml concentrated HNO_3 of Sp. Gr. 1.42 to 1 litre.

  Or

- Mixed reagent: Dissolve 25g ammonium molybdate and 1.25g ammonium metavanadate in 250 ml of warm distilled water separately in two beakers. Cool them and mix in a one litre volumetric flask, add 250 ml concentrated HNO_3, cool it and make up the volume with distilled water to one litre.
- Hydrochloric acid, approx. 5 M: Dilute 215 ml of Hydrochloric acid (approx, 36% m/m HC1) to 500 ml.
- Phosphorus stock standard solution, 1 mg/ml (or 1000 ppm) of phosphorus: Dry potassium dihydrogen orthophosphate at 102^0 for 1 hour and cool in a desiccator. Dissolve 0.879 g of the dried salt in water and add 1 ml of hydrochloric acid (approx. 36% m/m HC1). Dilute to 200 ml and add one drop of tolune to the solution.
- Phosphorus working standard solution, 0-50 4ug/ml of phosphorus — Prepare on the day of use solutions containing 0, 10, 20, 30, 40 and 50 ug/m1 of phosphorus.

Preparation of Standard Graph

Prepare 10 ml of each phosphorus working standard solution into a 50 ml volumetric flask. To each add 5 ml of approx. 5 M hydrochloric acid and 5 ml of ammonium molybdate-ammonium metavanadate reagent. Dilute to 50 ml and allow to stand for 30 minutes. Measure the absorbance in a 10 mm cuvette at 400 nm. Construct a graph relating absorbance to mg of phosphorus present. The absorbances corresponding to 0 and 50 ug of phosphorus are approx. 0 to 0.9 respectively.

Calculation

$$\text{P in plant material (ppm or mg kg}^{-1}\text{)} = \frac{\text{C x 50 x Volume of digest (100 ml)}}{\text{wt. of sample (g) x Aliquot (ml) taken}}$$

Wt. of plant sample = 0.5 g

Aliquot digest taken = 5 ml

Volume of digest 50 or 100 ml

$$\text{P in percent} = \frac{\text{C x 50 x Volume of digest (ml)}}{\text{Wt. of sample (g) X Aliquot (ml) X 10000}}$$

P in plant material (%) = P in plant material (ppm) x 10^{-4}

- The main advantage of the above method over molybdophosphoric blue colour method (used for available P in soil determination) is its lower sensitivity (1-20 ppm P) which permits its application on a macro scale and makes it more suited to plant analysis. Ions that do not interfere in concentration up to 1,000 ppm are : Al, Fe, B, Mg, Ca, Ba, Sr, Li, Na, K, NH_4, Cd, Mn, P1, Hg, Sn, Zn, Cu, Ni, Ag, U, Zr. OAC, As (ite), Br. CO_3, ClO_4 cynide, molybdate, pyrophosphate, silicate, nitrate, nitritie, sulphate and sulphite (Jackson, 1958). Extreme simplicity and colour stability ranging from 24 hours to 2 weeks make this method popular.

2.5. Determination of Potassium in Plant

Flame photometry method is most common method of K determination in plant. Sample can be digested in di-acid or tn-acid wet oxidation or dry ashing method as described earlier.

Principle

The determination is based on measurement of the spectral line intensities of potassium atom exited when passing through a flame. Atoms of some specific element like K takes energy from flame and get excited to the higher orbit. Such atoms release energy of a wavelength and give spectral lines which is specific for element and is proportional to the concentration of atoms of the element.

Apparatus

- Flame photometer
- Volumetric flasks

Reagents

- Potassium stock standard solution. 1 mg/ml (or 1000 ppm) of potassium: Dry potassium chloride at 102°C for 1 hour and cool in desiccator. Dissolve 0.953 g of the dried salt in water and add 1 ml of hydrochloric acid (approx. 36% m/m HCl). Dilute to 500 ml and add one drop of toluene.

- Potassium working standard solutions, 0-50 4ug/ml of potassium. Prepare solutions containing 0, 10, 20, 30, 40 and 50ug/m1 of potassium.

Preparation of standard graph

- Set the flame photometer, according to the manufacturers instruction to measure potassium emission.
- Nebulize the potassium working standard solutions containing 0 and 50 ug/ml of potassium and adjust the controls until steady zero and maximum readings are obtained. Nebulize the intermediate working standard solutions and construct a graph relating meter readings to ug/ ml of potassium in all the standard solutions.

Examination of sample solution

Measure 2.5 ml of the sample solution in to a 100 ml volumetric flask and dilute to 100 ml. Adjust the flame photometer until the steady zero and maximum readings are obtained with potassium working standard solutions containing 0 and 50 ug/ml of potassium, as described in the earlier paragraph (preparation of standard graph). Nebulize the diluted solution and note the meter reading.

Calculations

Wt. of Plant sample 0.5 g

Total Volume of Plant digest = 50 ml or 100 ml

R = Reading of flame photometer for sample - blank reading

$$\text{Potassium (ppm)} = \frac{\text{C(ppm) x Vol. of digest x 100}}{\text{Wt. of sample x 2.5}}$$

Where, C = Concentration of K read from standard curve

Potassium (%) = Potassium in plant material (ppm) X 10^{-4}

2.6. Estimation of Calcium and Magnesium in Plant by EDTA-Titration

(Source: Prasad, 1998)

Dry ashing

Take 0.1 to 1.0 g plant material in a porcelain or silica or platinum crucible, add drop by drop enough diluted alcoholic sulphuric acid solution (1% v/v concentrated H_2SO_4 solution in 95% ethanol) to moist the sample, and ignite the sample with the pipette (used for adding alcoholic sulphuric acid solution) torch to remove the excess alcohol. After the flame subsides transfer crucible to a muffle furnace and ignite at 525°C for 45 minutes or longer till grey-white ash free of carbon is obtained.

Extraction of Calcium and Magnesium

Add 5 ml of 2 N HCl to the crucible, let it stand for 15 minutes and then transfer the contents to a 100 ml volumetric flask, make up the volume and filter or centrifuge the solution.

Principle

A complex metric titration using EDTA is a classical method for determining Ca and Mg simultaneously or individually. Analytical success is based on eliminating possible interfering ions, achieving the required pH, and employing the appropriate indicator. Various ions such as Fe, Mn, Cu, Zn, Ni and phosphate may interfere with the chelation of the target ion by the EDTA molecule. This chelating reaction by EDTA is necessary for successful analysis. Cyanide, which forms very stable complexes with Cu, Ni and Fe may be added to eliminate their competition in the reaction.

Calcium

Reagents

1. 0.01 N standard Ca solution: Weigh 0. 5005g calcium carbonate dried at 150°C into 1 litre volumetric flask. Add 200 ml of 1 M HCI slowly with shaking. Make up the solution to 1 litre.
2. 0.01 N EDTA solution: Dissolve 0 .7306 g EDTA in about 200 ml water and make up the volume to 1 litre.
3. 2% NaCN or KCN solution : Dissolve 20 g NaCN or KCN in 1 litre of distilled water.
4. Muroxide indicator: Mix well 0.2 g Murexide (ammonium purpurate) and 40 g potassium sulphate.
5. 2 N sodium hydroxide solution: Dissolve 80 g NaOH in 200 ml water and make up to 1 litre.

Procedure

Transfer 25 ml extract to a porcelain dish, add 5 drops NaCN solution, mix well with a policeman tipped glass rod and add 10 ml of 2*N* NaOH solution. Now add a pinch of Murexide indicator and titrate with EDTA solution. When all Ca present in chelated by EDTA, the colour changes to purple.

Since the colour change takes place slowly and the detection of end point is rather difficult, it is always better to keep a standard purple colour obtained by titrating standard Ca solution with EDTA solution of known strength. Carry out a reagent blank.

Calculation

$$\% \text{ Ca in plant sample} = \frac{(\text{ml EDTA for sample} - \text{Blank}) \times \text{N* EDTA} \times 20 \times 4 \times 100}{1000 \times \text{Sample weight in (g)}}$$

N*EDTA = Normality of EDTA

MAGNESIUM

Magnesium is determined by first estimating Ca + Mg by titration with EDTA using Eriochrome black triethanolamine (EBT) and then subtracting the value of Ca (as estimated above) from it. Calculations are similar to as given above for calcium.

2.7. Estimation of Sulphur in Plant

Determination of total sulphur in plant material can be done by dry ashing or di-acid wet oxidation method, while tri-acid wet digestion is not suitable or applicable.

Principle

Total sulphur present in plant material is oxidised to sulphate during digestion process. The concentration of sulphate in solution is determined turbidimetrically, Barium sulphate ($BaSO_4$) is precipitated by addition of Barium chloride ($BaCl_2$) under standardized conditions. this provide turbidity to the solution which is proportional to the amount of sulphate present in solution. Measurement of this turbidity gives in sulphur content of solution.

Preparation of sample solution by dry ashing

Apparatus

1. Evaporating basins —50 ml glazed porcelain, shallow form with round bottom and spart.

Reagents

2. Hydrochloric acid, approx. 36% m/m HC1.
3. Magnesium nitrate solution — Dissolve 713 g of magnesium nitrate hexahydrate, in water and dilute to 1 lt.

Procedure

Transfer 1 g of dried sample, ground to pass 1 mm mesh sieve, into an evaporating basin. Add 10 ml of magnesium nitrate solution dropwise over the complete surface of the sample. Place the basin on a hot plate maintained at 180°C and when the residue is dry, raise the temperature of the hot plate to 280°C. When the residue colour changes from brown to yellow, place the basin in muffle furnace at 450°C and maintain this temperature (usually over night) until a white ash remains. Remove the basin from the muffle furnace, cool and cover with a watch glass. Moisture the residue with water, carefully add 10 ml pf hydrochloric acid and gently boil for 2 min. Add approx. 10 ml of water, remove and rinse the watch glass, collecting the washings in the basin. Filter the contents of the basin through a Whatman no. 41 filter paper into a 100 ml. Carry out a blank determination using 1 g of sucrose in place of sample.

Note: Since dry ashing leads to some volatilization loss of sulphur present in the organic combination and wet oxidation based on tri-acid mixture includes H_2SO_4. Thus, HNO_3 — $HC1O_4$ digest of plant is conveniently used.

Preparation of sample solution by di-acid digestion : (previous exercise)

2.8. Determination of Fe, Zn, Mn and Cu in Plant

These elements can be determined with the help of Atomic Absorption Spectrophotometer (AAS). Sample should be digested either by dry ashing or by di-acid oxidation method. Digestion of plant sample with H_2SO_4 or H_2O_2 is avoided because it contributes some micronutrient or heavy metals.

Principle

The atoms of metallic elements like Fe, Mn, Cu, Zn etc. absorb energy with subject to radiation of specific wavelength. The absorption of radiation is proportional to the concentration of atoms of the elements present in sample.

Apparatus

i. Atomic Absorption Spectrophotometer (AAS)

ii. Hollow Cathode lamps of Zn, Fe, Mn and Cu

iii. Hot plate with temperature control system

vi. Polyethylene bottel 100 ml capacity

v. Conical flask measuring cylinders 25m1, funnels etc.

vi. Whatman filter paper No. 1.

2.8.1 Iron (Fe)

Sample preparation : 1 g oven dried plant sample is digested using diacid mixture as described under wet digestion of plant samples and made up to 100 ml using deionized water on cooling.

Stock Standard Solution

Dissolve 1.0 g of pure iron wire in 50 ml of (1 + 1) analytical grade of HNO_3. Dilute to 1 lit, with deionized water, or (II) Dissolve 7.022 g of analytical grade of $(NH_4)_2$ Fe (SO_4) $6H_2O$ in 400 ml deionized water. Add 5 ml of conc. H_2SO_4 and make up the volume with deionized water. This would give a stock solution of 1000 ppm Fe. From this, solution of 100 ppm is prepared which will be used for preparation of final standard solution.

2.8.2 Manganese*(Mn)*

The procedure is similar to that of iron estimation.

Stock Standard Solution

Dissolve 1.0 g of manganese metal in a minimum volume of (1 + 1) HNO_3. Dilute to 1 lit. with 1% (V/V) HCl. Or (ii) Dissolve 3.076 g of $MnSO_4 . H_2O$ in deionized water and make the volume upto 1000 ml. This will give a solution of 1000 ppm Mn. 10 ml of this solution is diluted to 100 ml to get a solution of 100 ppm Mn. This solution is used for preparation of final standards.

2.8.3 Zinc(Zn)

The procedure is similar to that for iron estimation.

Stock standard solution

Dissolve 0.500g of zinc metal in a minimum volume of (1 + 1) HC1 and dilute to 11 with 1% (V/V) HC1 Or (ii) Dissolve 3.398 g of $ZnSO_4$ $7H_2O$ in distilled water and volume is made upto 1000 ml. This gives a solution of 1000 ppm zinc. 10 ml of this solution is diluted to 100 ml to get 100 ppm Zn solution. The final standard solutions are prepared from this'100 ppm solution.

2.8.4 Copper (Cu)

The procedure is similar to that for iron estimation.

Stock standard solution

Dissolve 1.000 g of copper metal in minimum volume of (1 + 1) HNO_3. Dilute to 1 lt with 1% (V/V) HNO_3Or (ii) Dissolve 3.929 g $CuSO_4$ $5H_2O$ in distilled water and volume is made up to 1000 ml. This gives a solution of 1000 ppm Cu. 10 ml of this solution is diluted to 100 ml to get 100 ppm solution of Cu.

Standard curve and calibration

Usually at least 5 to 7 standards and a blank of each micro-nutrient cation are used for drawing a calibration curve. The blank solution (0 mg/ml) is set at zero of the scale on the atomic absorption spectrophotometer. The Standards are then analyzed with lowest concentration first, and the blank run between standards to ensure that the base line (zero point) has not been changed. A graph of absorbance vs concentration of standard solutions is plotted on a graph paper. After calibrating the instrument, aspirate the sample and note the absorbance to find out the concentration from the conc. vs absorbance curve. Suppose it is M. Concentration of element (ug/g or ppm) M X 2

Calculation

Element in sample (Fe, Zn, Cu, Mn) = AAS reading (mg l^{-1}) X dilution

$$\text{Where, dilution} = \frac{\text{Final volume (ml)}}{\text{wt. of Plant tissue (g)}}$$

Or

$$= \frac{\text{R x 100}}{\text{wt. of sample}}$$

Similar calculation is followed for each element Zn, Fe, Mn and Cu.

Precautions

- Most of modern AAS are calibrated to display the concentration in mg l^{-1} directly. If AAS displays the reading in absorbance then a standard curve has to be prepared and reading should be converted into concentration (mg l^{-1} or ppm) from the curve.

2. If concentration is too low make final volume of dry ash digest upto 10 ml, in case of di-acid make final volume 20 ml.
3. AAS has greater sensitivity and accuracy.

4. To avoid contamination which is serious problem in analysis of trace elements, proper cleaning of glassware and lab. wares and running a blank solution is necessary.

2.9. Estimation of Boron in Plant

Principle

Total boron (B) in plant material is determined by dry ashing in presence of Calcium hydroxide [$Ca(OH)_2$] to prevent the loss of boron, the residue is dissolved in hydrochloric acid. The concentration of boron is measured with spectrophotometer as the yellow coloured complex formed with azomethine — H.

Apparatus

1. Evaporating basins 20 ml translucent silica, shallow form, with round bottom and spout.
2. Muffle furnace
3. Polythene tubes 15 ml
4. Volumetric flask 25 & 100 ml.

Reagents

1. Buffer masking reagent—Dissolve 250 g of of ammonium acetate and 15 g of ethylene-di-amine tetra acetic acid, sodium salt, in 400 ml of water. Carefully add 125 ml of acetic acid, glacial.
2. Calcium hydroxide solution, saturated.
3. Hydrochloric acid, M-Dilute 85 ml of hydrochloric acid, approx. 36% m/m HC1 to 1 litre with water.
4. Sucrose.
5. Azomethine-H reagent-Dissolve 0.45 g of Azomethine-H in 100 ml of 1% m/v L-ascorbic acid solution.
6. Boron stock standard solution (1000 ug B ml^{-1}). Dissolve 5.716 g of boric acid (H_3BO_3) in a small volume of deionized water and make final volume of 1 lit: This will have 1000 ug B ml^{-1}.
7. Working standard solution of B (20 ug B ml^{-1}) Dilute the stock solution by 50 times (take 2 ml and make it to 100 ml with deionized water).

Preparation of standard graph

Pipette 5 ml of each boron working standard solution into silica evaporating basin. Add 0.25 g of sucrose and 2 ml of saturated calcium hydroxide solution. Evaporate to dryness on a boiling water bath. Place the basin in a cold muffle furnace and slowly increase the temperature to 450°C. Maintain this temperature for 2 hours. Allow to cool, add exactly 5 ml of M hydrochloric acid and dissolve all soluble material. Filter through a 90 mm Whatman 41 filter paper. Transfer 1 ml of the filtrate to a polythene tube. Add 2 ml of buffer masking reagent, mix and add 2 ml of azomethine-H reagent. Mix well and allow to stand for 45 minutes. Measure the absorbance in a 10 mm optical cell at 420 mm. Construct a graph relating to absorbance to mg boron present. The absorbance corresponding to 0 and 3 mg of boron are approx. 0.2 and 0.7. respectively and should differ by approx. 0.45.

Procedure

Transfer 0.5 g of sample, ground to pass a 1 mm sieve, into a silica evaporating basin. Add 2 ml of saturated calcium hydroxide solution. Continue as in preparation of standard graph commencing at evaporate to dryness and ending at in a 10 mm optical cell at 420 mm.

Calculation

Read from the standard graph the number of mg of boron equivalent to the absorbance of the sample and multiply by 50. The result gives the μg kg^{-1} of boron in the sample.

wt. of sample 0.5 g

Absorption reading of spectrophotometer = A

Concentration of B as read from graph against A = C ug ml^{-1}

Dilution =50 times.

Final volume made after extraction of B in acid = 25 ml.

Content of B in plant sample (ppm or ug kg^{-1}) = C x 50

Notes: (a) Borosilicate glassware must not be used at any stage of the method. (b) Store all reagents in polythene bottle.

2.10. Estimation of Molybdenum in Plant

For the determination of total Mo in plants, sample is first digested with HNO_3 and $HC1O_4$ as described by Purivis and Peterson (1954) and Mo in the digest is estimated by colorimetric method of Johnson and Arkley (1954).

Principle

Molybdenum forms an amber orange coloured Mo complex when react with thiocyanate. This complex is extracted into an organic solvent and the concentration is determining through colorimeter.

Apparatus

1. Analytical balance
2. Conical flasks; 150 mL capacity.
3. Hot plate
4. Pipettes. 10 mL, graduated
5. Volumetric flasks: 100 mL capacity
6. Funnels
7. Filter Paper, Whatman No. 1

Reagents

1. Concentrated HNO_3 (AR Grade)
2. Perchloric acid
3. Hydrogen peroxide (H_2O_2)
4. Isopropyl ether: The ether is washed before use with a wash solution comprising a mixture of one-third stannous chloride ($SnCl_2$), one-third potassium thiocyanate (KSCN) and one-third deionized water. For washing, place reagent-grade isopropyl ether in a separatory funnel and add wash solution equal to one tenth of the volume of ether taken. Shake thoroughly and let organic phase separate from aqueous phase. Drain and discard the aqueous phase. Wash with 2N HCl again taking one tenth of the volume of ether. Shake and discard the aqueous phase. Repeat this process four to five times.
5. Hydrochloric ferric chloride: Dissolve 0.5g of hexahydrate ferric chloride ($FeCl_3 . 6H_2O$) to 560 mL of concentrated HC1 and dilute to one Litre with deionize water. This solution is required for preparing standards and reagent blank.
6. Potassium thiocyanate (40% w/v): Dissolve 40g of potassium thiocyanate (KSCN) in water and make up the volume to 100 mL with deionized water.

7. Stannous chloride (40% wlv) : Suspend 40 g of stannous chloride dihydrate ($SnC1_2 . 2H_2O$) in 20 ml of 6.5 N HC1. Add dionized or double distilled water to dissolve it and make upto 100 ml. Filter if turbid.
8. Stock Standard solution of Mo (100 u Mo/mL) : Dissolve 75 mg of pure molybdenum trioxide (MoO_3) in 5 ml of 0.1 N sodium hydroxide and dilute to little less than 500 ml. Make slightly acidic with HCI and make to 500 ml with deionized or double distilled water.
9. Working standard solution of Mo (1 ug Mo/mL) : Dilute 10 ml of the stock standard solution of Mo to one litre to obtain a working standard solution having 1 ug Mo/mL.
10. Citric acid (50%) : Dissolve 50 g of citric acid crystals in deionized or double distilled water and dilute to 100 mL.

Procedure

Digestion of plant samples

1. Weight 1-2 g plant sample and transfer it to a 150 mL conical flasks.
2. Add 15 mL di-acid mixture (prepared by mixing HNO_3 and $HC1O_4$ in 4: 1 ratio) and place a funnel at the mouth of the flask. Preferably keep the contents of the flask overnight.
3. Digest at low heat and raise the temperature slowly till white fumes of $HC1O_4$ start appearing. Evaporate the contents of the flask to dryness.
4. Cool and add 5 mL of HNO_3—$HC1O_4$ mixture and again evaporate to dryness.
5. Then add 1 mL H_2O_2 and allow the contents to dry and cool.
6. Add 50 ml of deionized or double distilled water, boil for one minute and add 10 mL concentrated HCl.
7. Make the final volume to 100 ml with deionized water.
8. Filter the solution and analyse it for Mo content.

Analysis of the digest

1. Transfer 50 mL of the digest to a separatory funnel.
2. Add 2 ml of 50% citric acid solution and dilute the contents to 60 mL with deionized water.
3. Add 2 mL of di-isopropyl ether and shake the flask for 2 minutes releasing pressure as necessary.

4. Discard the organic phase.
5. Add 15 mL of 40% potassium thiocynate solution and 15 ml of 40 percent stannous chloride solution.
6. Molybdenum in the sample will be converted into molybdenum thiocyanate with orange colour which should be extracted in 5 mL of di-isopropyl ether by shaking the contents of the separatory funnel.
7. Discard the aqueous phase and read the absorbance or per cent transmittance of the organic phase with a colorimeter at a wavelength of 470 nm.

Preparation of Standard Curve

1. Pipette 0, 0.5, 1, 2, 4 and 6 mL of solution containing 1 ug Mo/ml. in separatory funnels.
2. Add 10 mL of 6.5 N HCI-$FeCl_3$ solution to each separatory funnel.
3. Dilute the HC1-molybdate solution to 45 mL with deionized water.
4. Add 2 mL of 50% citric acid solution and dilute the content to 60 mL.
5. Add 2 mL of di-isopropyl ether and shake vigorously for 2 minutes, releasing pressure as necessary.
6. Allow 10 to 15 minutes for phase separation and discard the organic phase.
7. Add 15 mL of 40% KSCN solution and then 15 mL of 40 stannous chloride solution and mix again.
8. Add 10 mL of di-isopropyl ether and shake well.
9. Let the organic phase separate and then discard the aqueous phase.
10. Transfer the organic phase to the tube of the colorimeter and measure the transmittance or absorbance of the solution at 470 mm.

Calculation

Weight of plant sample taken for digestion = 1.0 g.

Final volume made after digestion = 100 ml.

First dilution = 100 times.

Volume of the fitrate taken in the separatory funnel = 50 ml

Volume of the organic solvent used for extraction = 5 ml

Second dilution = 0.1 times

Total dilution = 10 times

Absorbance shown by spectrophotometer = A

Concentration of Mo observed from the standard curve against A = C ug/ml

Amount of Mo in the plant sample = C x 10 ug/g

3

Water Quality Analysis for Irrigation

The quality of irrigation water is a crucial factor for long-term soil productivity. Poor quality water if used for a long time can make the soil less productive or even barren, depending on the amount and type of constituents present in it and the texture of the soil in question. Low or marginally saline waters sometimes appear to stimulate crop growth because of the higher amounts of nutrient ions present. However, excess of the soluble salts in water leads to their accumulation in the surface layer particularly in fine-textured or poorly drained soils. Many areas in the country are facing a serious problem of not only scarcity of water, but also of its extremely poor quality. Tube well or well waters generally pose such problems more than canal waters, especially in arid and semiarid regions. It is, therefore, advisable to get the water tested for quality while installing a tube well instead of repenting later. Locating appropriate strata during boring for tube well is to be done by 'hit and trial' method causing huge investments, if the knowledge about surrounding localities is lacking In such a situation samples of water should be tested while installation is in progress and the appropriate depth decided accordingly. Irrespective of the source of irrigation water, all that is required to be done is to draw a proper sample for testing in the laboratory.

3.1. Water Sampling Techniques

Water samples for irrigation quality assessment are analyzed for chemical constituents. Therefore, utmost care should be exercised to avoid the possibility of any external contamination. The samples can be collected in glass or plastic bottles thoroughly cleaned and rinsed 3-4 times with the water to be tested. For collection of water sample, the method described below may be adopted:

1. Run the tube well/hand pump for about 15-20 minutes to drain out the water retained in the pipe. Wash the bottle repeatedly with the water before taking about 500 ml. of sample. The bottle should not be washed with detergent or soap.

2. If a new tube well installation is in progress, collect water samples at different depths at intervals of about 3-4 meters. Since the water in such a situation is always turbid with suspended impurities, the sample may

be collected as such without caring for turbidity, as it does not affect the test results. It is to be filtered before analysis any way.

3. If water from a tank or pond is to be tested, collect the sample from at least 5 to 10 meters away from the boundaries. For this purpose tie a suitable (small) container at one end of a long bamboo pole and collect the sample after displacing surface water, which might contain organic material floating over it.
4. From a well, the sample can be drawn either during irrigation, just before the water falls in the channel or by drawing it with the help of a bucket or any other clean container using a rope. The water surface should be disturbed a little to remove iny floating material before collection of the sample.
5. Place the cap on the bottle tightly.
6. Take a piece of paper of suitable size (about 8 cm x 4 cm), write the name, address, sample identification mark *etc.* and paste it firmly on the bottle.
7. Write separately on a sheet of paper the name of the crop to be grown, texture of soil to be irrigated, previous experience about the effect of the water on soil surface, if any, earlier test results if available, other sources available for irrigation, and crop performance so far. Send into the laboratory along with the sample, preferably within 2-3 days of collection of the sample.
8. In case any delay is anticipated in sending the sample for testing or the samples are to be stored for more than a week, add 2 to 3 drops of toluene to the sample to prevent microbial growth or preferably draw a fresh sample.
9. It is desirable to send both soil and water samples together for testing, as it helps in making rational recommendations.

3.2. Chemical Analysis of Water

3.2.1 pH of water

The determination of pH of water samples is based on the same principle as given for soil under exercise of soil p^H determination

Instrument

1. pH meter

Reagents

1. Buffer Solutions of pH 4.0, 7.0 and 9.2

Procedure

1. Standardize the pH meter with the three pH buffer solutions after giving warm up time, if needed.
2. Take about 50 mL of water sample in a 100 mL beaker (preferably plastic), immerse the electrodes of the pH meter and record the pH value.

3.2.2. Soluble Salts

Principle

The total soluble salts can be determined as the electrical conductivity C) of the water sample using a conductivity meter/solubridge as in case of soil-water suspension. The electrical conductivity is directly proportional to the amount of salts (ions) and is expressed as dSm^{-1} (which is equivalent to the unit mmhos cm^{-1}). Usually, a temperature of 20 °C is kept for the EC determinations.

Instrument

1. Conductivity Bridge (EC meter)

Reagent

1.0.01N KCI solution: Prepare as described under determination of EC of soil

Procedure

1. Switch-on the conductivity meter and allow warm up time, if required.
2. Standardize the EC meter with 0.01 N KCl which gives a value of 1.41 dSm^{-1} at 25 °C if the cell constant is 1.0.
3. Fill the conductivity cell or dip the cell assembly in water sample taken in a 50 or 100 mL beaker and record the conductivity. If the value is too low, change the position of the knob suitably.
4. Alternatively, evaporate a known volume of water and weigh the dry residue (gravimetric method). However, the conductivity method is more precise and rapid, specially useful when the quantity of water sample is limited and the same is to be used for further analysis.

Calculation

The EC value can either be used as such for categorizing the water on salinity basis or may be used to get the concentration as given below:

Total salt content (mg L^{-1}) = EC (dSm^{-1}) at 25 °C *x* 640

or

Total salt content in meL^{-1} (approx.) = EC (dSm^{-1}) at 25 °C X 10.

3.3. Calcium and Magnesium

Principle

The extent of sodium hazard in irrigation water is gauged in terms of the sodium concentration in relation to Ca^{2+} and Mg^{2+}. For the determination of both of them together, the versenate titration method is most popular in which EDTA- disodium salt solution is used to chelate them. The two cations can also be precisely estimated using AAS, but for all practical purposes the versenate method is good enough. Calcium alone can also be estimated by versenate method using ammonium purpurate (muroxide) indicator and thus Mg^{2+} can be obtained by subtracting Ca^{2+} from Ca^{2+} and Mg^{2+} content. Calcium estimation can also be done on a flame photometer but the precision is not very high. A number of cations including $Ca2^{+}$ and $Mg2^{+}$ are known to form stable complexes with versenate at different pH values. The formation of Ca and Mg complexes at pH 10 is achieved by using ammonium hydroxideammonium chloride buffer. A number of polyvalent ions are preferentially complexed by versenate as these are less dissociated than those of Ca and Mg and thus included in the titration. Fortunately the concentration of such interfering metals *e.g* Fe, Cu, Pb, Cd, Zn, Co, Mn is quite low and negligible in most waters and can be ignored. However, the interference, if high, can be prevented by using 2% solution of NaCN.

Reagents

1. Standard $CaCI_2$ solution (0.01 *N):* As given under section 1.3.1.
2. Standard versenate solution (0.01N): As given under section 1.3.1.
3. Ammonlum hydroxide-ammonium chloride buffer (pH 10): As given under Section 1.3.1.
4. Erlochrome black T (EBT) indicator: As given under section 1.3.1.

Procedure

1. Transfer 5 mL of the water sample to a 100 mL porcelain dish and add about 20 mL of distilled water.
2. Add 1 mL of ammonium hydroxide-ammonium chloride buffer and 3 to 4 drops of EBT indicator.
3. Take versenate solution in a burette and titrate the contents in the dish. The wine red colour changes to light/sky blue at the end point.
4. Swirl the contents in the dish to ensure complete disappearance of the red colour.

Calculation

Ca ± Mg concentration in water (me μ^{-1}) = 2V

where, V is the value of 0.01 NEDTA solution used (titre value) for 5

mL of water sample.

The factor '2' is arrived at from the expression given in exercise-3.4.

N1V1 = N2V2

where, N1 and N2 stand for the concentrations in me L1 for water and EDTA solution used, respectively, and V1 and V2 for the volumes of water (5 mL) and EDTA solution consumed (V).

(Normality of the EDTA solution is 0.01NLe. 10 me L-1)

3.4. Estimation of Carbonates and Bicarbonates

Principle

Carbonates and bicarbonates in water can be conveniently determined by titrating a known volume of water against standard H_2SO_4 using phenolphthalein and respectively. When a drop of phenolphthalein is added to water, the pink colour developed will indicate the presence of carbonates. In good quality water, generally pink colour will not appear, as carbonates are absent. On titration, the pink colour disappears because all the carbonates are converted to bicarbonates as shown below:

$2Na_2CO_3 + H_2SO_4 — 2NaHCO_3 + Na_2SO_4$

This neutralization of carbonates is only halfway. The bicarbonates thus obtained along with the already present ones are then determined by continuing the titration using methyl orange indicator which gives yellow colour in presence of bicarbonates. On complete neutralization of bicarbonates the yellow colour will change to red. The second stage neutralization takes place as follows: $2NaHCO_3+H_2SO_4—Na_2SO_4+2H_2O+2CO_2$

After completing the first titration (if pink colour appears) either the same aliquot is used for bicarbonate titration or a fresh sample is taken for it.

Reagents

1. Standard H_2SO_4(0.01 *N*): Carefully add 2.8 mL of conc. sulphuric acid to about 500 mL distilled water taken in a glass beaker and dilute to 1L to get approximately 0.1 NH_2SO_4. Dilute 100 mL of this solution to 1L. Standardize it by titrating against standard alkali prepared separately.
2. Phenolphthelein Indicator: Dissolve 0.25 g of phenolphthalein powder in 100 mL of 60% ethyl alcohol.
3. Methyl orange Indicator: Dissolve 0.50 g of the dye in100 mL of 95% ethyl alcohol.

Procedure

1. Transfer 5 mL of water sample to a 100 mL porcelain dish and add about 20 mL of distilled water.
2. Add 2 drops of phenolphthalein indicator.
3. If pink colour appears, titrate against 0.01 NH_2SO_4. Note volume of acid consumed (A mL).
4. Add 2 drops of methyl orange indicator to the colourless contents in the dish (left after the first titration or to a freshly taken 5 mL water sample). This will give yellow colour.
5. Continue the titration until the yellow colour changes to red. Note the volume of acid consumed in the second titration (B mL).

Calculation

Carbonates (me L1) = 4A, and

Bicarbonates (me L1) = 2 (B - A)

where, A and B stand for the volume of 0.01NH_2SO_4 (mL) consumed in the first and second titrations, respectively.

Note

The first titrate value (A) has been subtracted from the second (B), because (B) includes the volumes consumed for already present bicarbonates as well as for those obtained from neutralization of carbonates. The basis of the factor 2 is same as described for Ca ± Mg in water and it has been doubled for carbonates because only half neutra l iation is done in the first titration.

3.5. Determination of Sodium in Water Samples

Principle

Irrigation waters may have two major types of hazard *viz.* salinity and sodium hazard. The latter is expressed as the residual sodium carbonate (RSC) and sodium adsorption ratio (SAR). For working out the SAR value, sodium content is determined conveniently using flame photometer or simply by subtracting the Ca + Mg content from the total soluble salts. The subtraction method, though not so precise due to the presence of other canons like K, is often adopted to serve the purpose. The quantity of K ions in water is generally low enough to be ignored for this purpose. The flame photometric determination of sodium is given below:

Instrument

1. Flame photometer.

Reagents

1. **Standard stock solution (100 me Na L^{-1}):** Dissolve 5.845 g of AR grade dried NaCl in distilled water and make the volume to 1 L.
2. **Working standard solutions of Na:** Dilute 5, 10, 15,20, 30, 40 and 50 mL portions of the stock solution (containing 100 me Na L^{-1}) to 100 mL in volumetric flasks to get working standards of 5, 10, 15, 20, 30, 40 and 50 me Na L^{-1} concentrations.

Procedure

1. Filter a portion of the water sample if suspended material is visible. Filtration is desirable as it prevents choking of the capillary tube of the flame photometer.
2. Take the working standard solutions and record the flame photometer reading against each, after setting zero with distilled water and 100 with the highest concentration *i.e.* 50 me Na L^{-1}.

3. Feed the test sample and record the reading.
4. Draw a standard curve by plotting the readings against Na concentrations.

Calculation

Na concentration in water (me L^{-1}) is directly obtained from X-axis against reading.

Calculations for Sodium 1-Hazard Parameters

i) ***Residual Sodium Carbonate (RSC)*** The residual sodium carbonate may be calculated simply by subtracting the quantity of Ca + Mg from the sum total of carbonates and bicarbonates determined separately in a given sample and expressed in me/L. Thus,

$$RSC = (CO_3^{2-} + HCO_3^{1-}) - (Ca^{2+} + Mg^{2+})$$

ii). *Sodium Adsorption Ratio (SAR)*

Sodium adsorption ratio is calculated using the formula:

$$SAR = \frac{Na}{\sqrt{\frac{Ca^{2+} + Mg^{2+}}{2}}}$$

Determination of Potassium in water sample

Potassium in water can be easily determined by flame photometric method exactly in the same manner as sodium, except that the working standards of much lower concentrations (e.g 0,2,4,6, 8, 10 and 20 tg K mL^{-1}) are required. These concentrations should be divided by equivalent weight of potassium to convert them into me L^{-1}.

3.6. Determination of Chloride in Water Samples

Principle

Chlorides being highly soluble are present in all waters, but the amounts are often very low in natural waters. However, their contents may be appreciable when the electrical conductivity is high. Chlorides can be easily determined by $AgNO_3$ titration (Mohr's titration) method in which silver reacts with chloride forming white AgCl precipitate. When all the chlorides are precipitated, potassium chromate (the indicator used) shows the brick red colour at the end point due to the formation of silver chromate.

Reagents

1. **Standard sodium chloride solution (0.02 N):** Dry AR grade NaCl in oven at 80°C for 1 hour, and dissolve 1.170 g of it in chloride-free distilled water and make to 1 L.
2. **Standard silver nitrate solution (0.02N):** Dissolve 3.40 g of $AgNO_3$ in about 200 mL of freshly prepared chloride-free distilled water and make to IL. Standardize it against 0.02NNaCI solution. Store in an amber bottle, away from light.
3. **Potassium chromate Indicator:** Dissolve 5 gof pure K_2CrO_4 in 100 mL of distilled water.

Procedure

1. Transfer 5 mL of water sample to a 100 mL porcelain dish.
2. Add about 20 mL of distilled water and 5 to 6 drops of potassium chromate indicator.
3. Titratc the contents against 0.02N$AgNO_3$ solution till the brick red colour just starts appearing.

Calculation

Chloride content in me L^{-1} = 4V

where, V denotes the titrate value in mL

(1 mL of 0.02Nor 20 me L^{-1} $AgNO_3$ 0.00071 g of chloride)

Determination of Sulphate in water samples

The water sample has to be filtered using Whatman No.1 filter paper to remove physical impurities. If the sample is coloured, it should be decolorised with activated charcoal. Thereafter, turbidimetric procedure (Chesnin & Yien 1950) as described in section (determination available sulphur) can be adopted for determination of sulphate -S in water.

Determination of Boron in water samples

For irrigation quality rating, water samples are not usually tested for B content. However, saline waters of arid and semi-arid regions may have high concentrations of B, which could be toxic depending upon the type of minerals present in soil and the nature of soil strata through which the water passes to reach the groundwater. Thus, some water from entrapped underground sources may contain excessive amounts of boron, which is toxic to plants. In fact, the margin between toxicity and deficiency limits of boron is very narrow and hence water samples suspected to be containing high B contents should be tested for B and classified accordingly. Boron in water can be determined by azomethine H method as described for soil boron under section (available sulphur in soil).

3.7. Biochemical Oxygen Demand (BOD)

The biochemical oxygen demand (BOD) of sewage or effluent indicates the amount of organic compounds present in it. It is considered as a good index of organic pollution. High amount of organic matter in water leads to higher oxygen utilization by bacteria for degrading it. Use of sewage having high BOD increases the concentration of soluble organic compounds in the aquatic body where it is discharged. Digestion of these compounds in natural ecosystems can deplete available oxygen, thus leading to asphyxiation of fish. Fortunately, most of the ground waters and well or canal waters do not contain high organic matter.

Principle

The determination of BOD is based on the measurement of the volume of oxygen required by bacteria to metabolize the organic compounds. The BOD is generally measured by incubating the water sample at 20°C for five days in the dark in aerobic conditions. In the tropical and subtropical climate, where both temperature and rate of metabolic activities are high, the incubation is preferably done at 27 °C for three days. Since a significant amount of oxygen is required for nitrification, there is an over-estimation of BOD, if it is not checked by a suitable method, e.g by adding 1 mL of 0.5% solution of allyl thiourea.

Instruments

1. pH meter
2. BOD incubator

Reagents

1. **Allyl thiourea:** Prepare 0.5% solution.
2. **Hydrochloric acid:** Prepare approximately 1 N solution.
3. **Sodium hydroxide:** Prepare approximately 1 N solution.
4. **Sodium thiosulphate:** Prepare 0.02N aqueous solution (eq. wt. 248.18g)

5. **Manganous sulphate solution**
6. **Alkaline Iodide-azidesolution**
7. **Starch Indicator:** Prepare 1% aqueous solution.
8. **Conc. sulphuric acid**

Procedure

1. Neutralize the water sample using 1 *N* acid or 1 *N* alkali (attain a pH of 7.0).
2. Fill the water sample in 6 BOD bottles (250 mL) without bubbling.
3. Add 1 mL of allyl thiourea to each bottle.
4. Add 2 mL of manganous sulphate and alkaline iodide-azide solutions in succession to three of the six BOD bottles, right at the bottom with separate pipettes and replace the stopper.
5. Shake the bottles in the upside down direction at least six times.
6. Allow the brown precipitate to settle.
7. Add 2 mL of conc. H_2SO_4 and shake the stoppered bottle to dissolve the brown precipitate.
8. Take 50 mL aliquot in a 250 mL conical flask and titrate with sodium thiosulphate solution till the colour changes to pale straw.
9. Add 2 drops of starch solution to the flask. It will change the colour from pale to blue.
10. Titrate again till the blue colour disappears.
11. Calculate the dissolved oxygen in mg L^{-1} as given below.
12. Take the mean of the dissolved oxygen of the three bottle (D1).
13. Incubate the rest three BOD bottles at 27 °C in a BOD incubator for 3 days.
14. Estimate the oxygen concentration in all the three incubated samples and take mean (D_2).

Calculation

$$\text{Dissolved oxygen (mg L}^{-1}) = \frac{(8 \times 1000 \times N) \times t}{V}$$

where

V = volume of sample taken,

t = total titre value (mL) and

N = normality of titrant *i.e.* 0.025Ps.

BOD (mg L^{-1}) = D_1-D_2

where, D1 and D2 are the dissolved oxygen contents in the initial (unincubated) and the incubated samples, respectively.

3.8. Chemical Oxygen Demand (COD)

With the alarming increase in pollution through the discharge of large amounts of various chemically oxidizable organic substances entering the aquatic systems, the biochemical oxygen demand alone does not give a precise estimate of the organic load of a water sample. Besides this, the presence of various toxicants in the sample may severely affect the reliability of BOD results. Therefore, the chemical oxygen demand (COD) is considered as a better indicator of organic matter.

Principle

The organic matter content of water is estimated by their oxidability by using chemical oxidants like potassium permanganate or potassium dichromate. In the case of the former, a known volume of $KMnO_4$ is first used to oxidize the organic matter and then the excess of oxygen is allowed to react with potassium iodide. This liberates iodine in amounts equal to the excess oxygen. The liberated iodine is estimated by sodium thiosulphate titration using starch as indicator.

Apparatus

Water bath

Reagents

1. 0.1 *N* Potassium dichromate solution
2. 0.1 *N* Sodium thiosulphate solution
3. 2M Sulphuric acid
4. 1% Starch solution

Procedure

1. Transfer 50 mL water sample in triplicate in 100 mL conical flask.
2. Carry distilled water blank also (in triplicate).
3. Add 5 mL of $K_2Cr_2O_7$ solution to each flask.
4. Place the flasks in water bath at 100 °C for one hour.
5. Cool the flasks for 10 minutes and add 5 mL of KI solution and 10 mL of H_2SO_4 to each flask.
6. Titrate the contents against 0.1 M sodium thiosulphate until pale yellow colour appears.
7. Add 1 mL of starch solution (it gives blue colour).
8. Titrate again until blue colour disappears completely.
9. Find out the total volume of titrant consumed and take the mean of the three flasks of each sample.
10. Find out the total volume of the titrant used for the blank also.

Calculation

$$\text{COD (mg L}^{-1}\text{)} = \frac{8 \times C \times (B-A)}{V}$$

where

C = concentration of titrant (m mol L^{-1}),

A = titrate value for blank,

B = titrate value for sample, and

V = volume of water sample taken.

3.9. Use of Saline/Sodic or Brackish Waters for Irrigation

As described above, water can be safely used for irrigation as long as the sum total of all the indices and weightage for soil texture does not exceed 9. Highly saline waters, however, can not be recommended for all crop categories either due to high value of residual sodium carbonate or high sodium adsorption ratio. If another source of good quality irrigation water is available, the high SAR water can be used after diluting it suitably. Dilution does not strictly mean the mixing of the two sources in a particular ratio but the aim could be achieved by alternate irrigations as well. Dilution in more than 1:2 ratio is generally not recommended obviously because such high SAR waters may as such be dropped and the better source alone used for irrigation. In case the availability of good water is very limited, the saline water may be used at a later stage of plant growth while the germination and early growth stages may be managed with safe water.

Waters having high amounts of residual sodium carbonate (RSC) may also be used, but along with the required quantity of gypsum to neutralize the excess RSC. For all practical purposes, gypsum may be recommended for soil application @ 40 kg/irrigation ha for each milliequivalent of RSC to be neutralized.

Interpretation of water analysis data

Various workers have suggested the criteria of classification of irrigation waters on the basis of analysis made (Richards 1954; Paliwal and Yadav 1976; Ayers and Westcot 1976). Some important and widely acceptable ratings are given below. These should be taken as a general guideline and necessary correction may be made depending upon the soil-crop situation.

Parameter	Permissible	Moderately safe	Moderately unsafe	Unsafe
RSC (me l^{-1})	<1.25	—	1.25 — 2.50	>2.50
SAR	<10	10—18	18—26	>26
Boron (mg l^{-1})	<2.0	2.0 — 2.5	2.5— 3.0	>3.0
Chloride (mg l^{-1})	<140	140 — 350	—	>350
BOD (mg l^{-1})	$\leq$ 100	-	-	>100

4-Instruments Used In Soil, Plant and Water Analysis

Most of the analytical methods used in a soil testing laboratory for soil, plant and water analysis involve some kind of instrument. Although some tests are conventionally carried out using simple titrimetric or gravimetric techniques, use of common as well as sophisticated instruments undoubtedly increase the precision of the results and speed- up analytical work. This makes the analysis cost-effective too. Based on usage, ease of handling and other associated factors, the instruments are broadly categorized into following groups:

i) General purpose instruments.

ii) Advanced instruments used for specific tests and data processing.

A pH meter, conductivity meter, colorimeter or spectrophotometer and flame photometer are placed in the first group while atomic absorption spectrophotometer (AAS), autoanalyser, inductively coupled plasma atomic emission spectrophotometer (ICP-AES) *etc.* are included in the second.

It has been generally realised that the lack of knowledge about the basic instrumentation principles, major components and their functions, operation and maintenance not only affects the test results and their accuracy but also the service life of instruments. Therefore, attempts have been made in this chapter to briefly explain these aspects so as to permit the best use of the instruments, which are often abandoned on account of even minor faults.

4

General Purpose Instruments

4.1. pH meter

A pH meter is very commonly used for the measurement of the reaction of a solution or a medium. The pH determination with the help of a pH meter is based on the measurement of the electromotive force (e.m.f) of a pH cell consisting of an electrode sensitive to W ions and a reference electrode. The pH cell can be represented as follows:

H^+ sensitive	Reference buffer or	Salt bridge	Reference
electrode	test solution electrode		

The conventional type of pH meters have a H^+ sensitive glass electrode and a calomel reference electrode. The other types of H^+ sensitive electrodes are hydrogen gas electrode, platinum electrode, quinhydrone electrode and antimony electrode.

Glass Electrode

The electrode is a glass membrane electrode. It consists of a H responsive thin-walled soft glass bulb sealed to a stem of non-H^+ responsive high resistance glass. The pH response is confined to the area of the glass membrane only and thus the effect of depth of immersion is eliminated during the pH measurement. Both the inner and outer surfaces of the bulb are sensitive to the changes in H^+ concentration. The inside of the bulb is filled with an electrolyte of high buffering capacity such as a 0.025Msolution of HCl or H_2SO_4. An inner reference electrode also clips into this electrolyte. When the glass electrode bulb is dipped in a solution having H^+ concentration different from that inside the electrode, an electrical potential develops across the membrane and this potential is proportional to the difference in the pH between the two sides. Since the pH of the inner side remains unchanged, the difference denotes the H^+ concentration of the outer solution. The glass electrode offers many advantages such as:

1. The dissolved CO_2 and other gases are not expelled from the system,
2. It is adaptable to thick fluids, pastes or gels,

3. There is no effect of oxidizing or reducing solutions,
4. No H_2 gas or catalytic surface or auxiliary materials are required,
5. It provides rapid and fairly accurate results, and
6. It is convenient for continuous working.

However, the glass membrane attains the equilibrium slowly and normally takes several seconds for it. Also, the glass surface is easily contaminated with adsorbed cations and suspended particulates, leading to delayed equilibrium. For an appropriate combination of chemical durability and electrical resistance the electrodes are designed for certain ranges of both temperature and pH. A general purpose electrode is useful as it works satisfactorily for —5 to 60°C with negligible error up to pH 11. The error at higher pH is considerable and needs correction.

Calomel Electrode

The reference calomel electrode consists of Hg in contact with KCI saturated with Hg2 C12. Undissolved KC1 salt in the saturated KC1 solution is clearly visible. In a number of commercial makes, a paste of Hg and Hg2 C12 is filled in the inner tube connected to the KC1 solution in an outer tube. The lead wire is connected to the paste through the Hg column. The outer tube containing KC1 ends in a fine capillary to act as a salt bridge through the test solution to the H sensitive electrode.

Measurement of pH

The term pH was introduced by Sorensen and is defined as the negative logarithm of the H activity, which is expressed in gram ions per liter. The H activity in a very dilute solution can be expressed as concentration in gram mole per liter. Thus,

$pH = -\log_{10} {}^{a}H^{+} \text{or} -\log_{10}[H^{+}]$

The effective concentration of H* ions includes all sources *e.g.* H obtained from dissociation of soluble acids and from soil particles.

Water dissociates into its components namely H and OH- ions but only slightly so much so that the quantity of the undissociated water remains large enough to be treated as constant. Therefore, in the ionization equilibrium:

$H_2O—H^+ + OH^-$

the dissociation constant K is expressed as:

$$K = \frac{[H^+] \times [OH^-]}{[H_2O]}$$

The product of the two ions *ie.* H^+ and OH- is also constant at a given temperature.

Thus,

$[H^+] \times [OH^-] = 10^{-14}$ g mol L^{-1}

At neutral pH (7.0), $[H^+] = [OH^-] = 10^{-7}$g mol L^{-1}.

Since the potential of glass electrode is directly proportional to pH, the scale is calibrated into pH units. The value of the potential at 25 °C is 0.059 volt for each pH unit. The activity of $H^+$and OH^- ions at different pH values on the scale are given below:

$$\frac{pH\ [H^+]\ [OH^-]}{0}$$

The electromotive force (e.m.f.) measurement is done on a suitably calibrated galvanometer or ammeter and read directly in pH units. It is necessary to set the instrument at the temperature of the test solution.

Care in Handling and Maintenance

The most delicate part of a pH meter is the glass electrode. The thin-walled glass bulb very often breaks due to rough handling. It may also dry up when kept out of water for long. If dried, the electrode should be immersed first in 0.1NHC1 and then in distilled water for one or two days and checked again for its sensitivity. For general care, attention may be paid to some important tips which are as follows:

1. For checking the defect in electrode, short circuit the electrode jacks with the special strap and provide the terminal connections with the instrument. Switch on the instrument and allow 15-20 minutes for warm up. If the galvanometer reads 7.0 or can be adjusted at pH 7.0, the defect lies in electrode. Replace the electrode in such a case.
2. Ensure that there is always some solid KC1 in the reference (calomel) electrode.
3. Remove the fine layer of clay gels, deposited on the electrode, if any, with the help of cotton wool and distilled water.

4. Keep a pair of glass and calomel electrode (or a combined electrode) as spare for emergency use.

4.2. Conductivity Meter

Principle

Salinity in soil and water is characterized as the total content of the dissolved inorganic solute. It is conventionally determined by measuring the electrical conductivity The electrical conductivity is a measure of the ability of a salt solution to carry electric current by the migration of ions under the influence of an electric field, as ions are the carrier of electricity. Like a metallic conductor, solutions also obey Ohm's law. Increase in temperature promotes dissociation of the salts with a consequent rise in conductivity at the rate of approximately 2% for each degree celsius rise in temperature (Bower and Wilcox 1965). The unit of specific conductance is the reciprocal of specific resistance in ohms cm, *i.e.* mhos cm1 If Cs is the concentration of a solution in gram equivalents L', then the volume of solution in mL per equivalent is 1000/Cs, so that conductance = 1000 K/Cs, where, K is the cell constant. At infinite dilution, the ions are theoretically independent of each other and each ion has its contribution to the total conductance.

Thus, where, ACs is the total conductance, is the conductance of cations, and

A is the conductance of anions at infinite dilution

The instrument used for measuring conductivity is also known as conductivity bridge. A typical system consists of an alternating current (AC) Wheatstone bridge, a primary element of conductivity cell and a null balance indicator (as in "Solubridge") or an electronic "eye" in the conductivity meter. The test solution is filled in a conductivity cell, which is usually made out from two platinum sheets embedded in glass so that the two surfaces facing each other remain exposed. The area and distance between these plates must remain constant. In order to increase the sensitivity of the measurement, one or two stages of amplification are provided before feeding the signal to the electronic eye.

Determination of Cell Constant

The cell constant, k, of a conductivity cell is determined by measuring the electrical conductance of a standard KC1 solution at the same temperature according to the equation:

$$K=L/C$$

where, L and C are the known specific electrical conductance and the measured conductance of the standard solution (usually 0.02MKC1) in the given cell in dS m^{-1}. The specific conductance (L) of 0.02MKC1 is 2.39 dS m^{-1}. at 18 °C and 2.77 dSm-1 at 25 °C. By using the temperature coefficient (Table 4.1) the specific conductance can be calculated at various temperatures. The measured conductance C, of a test solution in dS m^{-1}. multiplied by the cell constant (k) gives the specific conductance, L in dS m^{-1}. of the test solution (L = kC).

Procedure

1. Switch on the instrument. Allow at least 15-20 minutes for warm up.
2. Wash the electrode with distilled water and wipe dry with tissue paper.
3. Use 0.02MKC1 solution to calibrate the meter.
4. Wash and fill the cell with the KC1 solution. Adjust the meter to read the standard conductivity at that temperature or adjust the cell constant with the help of calibration knob.
5. Wash the electrode with distilled water, followed by the test Solution.
6. Fill the electrode with the test solution.
7. Record the conductivity reading. Note the temperature of the solution and make the temperature correction (Table 4.1) to express results as dS m^{-1}. at 25 °C.

Precautions

1. Before using a new conductivity cell, place it in distilled water for 24 hours.
2. When not in use, keep the cell in distilled water.
3. Keep the cells perfectly clean for accurate readings.
4. Rinse the cell with the specimen before making the measurement.

Table 4.1: Conversion factors for obtaining EC values at 25°C

Observation Temperature (°C)	Conversion Factor	Observation Temp. (°C)	Conversion Factor
14.0	1.217	20.2	1.107
15.0	1.247	20.4	1.102
16.0	1.213	20.6	1.097

17.0	1.189	20.8	1.092
18.0	1.163	21.0	1.087
18.2	1.157	21.2	1.082
18.4	1.152	21.4	1.078
18.6	1.147	21.6	1.073
18.8	1.142	21.8	1.068
19.0	1.136	22.0	1.064
19.2	1.131	22.4	1.060
19.4	1.127	22.6	1.051
19.6	1.122	22.8	1.047
19.8	1.117	23.0	1.043
20.0	1.112	24.0	1.020

4.3. Flame Photometer

Principle of Flame Photometer

The conventional methods of analysis are confined to either gravimetric (where a precipitate is obtained and weighed) or volumetric procedures which include neutralization titrations, oxidation-reduction titrations, precipitation titrations and complexmetric titrations. Sodium in soils can be estimated gravimetrically by precipitation as sodium uranyl zinc acetate whereas potassium can be determined as potassium cobaltinitrite. Due to the lengthy and tedious procedures involved, resist is sought in more time saving and convenient procedures. One of these methods is the Flame photometry. In Flame photometry the spectral light emitted by substances at elevated temperatures is quantified in terms of light energy measurement. Essentially a Flame photometer consists of three parts:

(a) A source to excite the atoms;

(b) A source to isolate the particular radiation desired, and;

(c) A light measuring source.

A flame provides the first source usually oxygen acetylene. In the instrument, we have in the laboratory the source is methane and compressed air. The second source is fulfilled by eihter a peism or grating. in our instrument we use filter. for measuring the light emited, a photocell Is used which is connected to a galvanometer, which gives the relative readings. From readings known by atomizing different concentrations of a particular element, the concentration in the unknown liquid is determined.

When a solution containing Na or K (or Ca, Mg, Ba, Sr and other cations) is placed in the nozzle below the flame, the solution is sprayed into the flame in form of a fine spray. The high temperature in the flame evaporates the liquid whereas the solid present is excited. During excitation, the electrons present in the orbit closest to the nucleus gain energy and jump to the next higher orbit. But in the next higher orbit, due to the number of electrons which can be there being fixed ($2n^2$, where n is equal to the number of the orbit) the electrons jump to the original orbit. During this process energy is given off which is measured by the photocell. We can represent it as:

$AE = E_2 - E_1 = nf$

Whereas

AE = Change in energy

E_2 = Energy level in higher orbit (excited)

E_1 = Energy level in lower orbit (normal state)

n = Planck's constant (66234 X 1027 erg sec1) and

f = Frequency of light

The energy E radiated depends on the number of excited atoms by the flame, which in turn depends on the concentration of the particular element in solution. Source: Prasad et al. (2006)

4.4.Atomic Absorption Spectrometer (AAS)

Principles of Atomic Absorption Spectrometer (AAS)

Atomic absorption, as the name implies, depends on the measurement of atomic species. The whole technique depends on the ability to consistently atomize the element of interest in a reproducible manner in both samples and standards.

Atoms can be excited by thermal or other means to higher energy states (atomic absorption), and they then emit radiation by dropping down to less energetic states or to the ground state (emission). The process can be summarized as follows:

$M.+hv \text{---}> M^* \text{---}> M+hv$

Adsorption Emission

Where M is the neutral atom, M^* & is the excited atom, h is Planck's constant, and v is the frequency. (The difference in energy, between the two states E^* — E is equal to AE = hv = hc a^, where c is the speed of light and a the wavelength).

In absorption, the energy imparted to the atom (hv) is derived from the light source, whereas in emission this energy is primarily imparted to the atom in the atomization-excitation source.

The number of atoms in a flame in a particular excited energy state (n^*) and the number of atoms in the ground or unexcited state (n) are related by the following familiar Boltzman relationship:

$n^*/n = A \exp(-AE/KT)$

Where A is proportionality constant involving the ratio of statistical weights of the high and low Energy State, K is the Boltzmann constant, and T is the absolute temperature. Typical n^*/n values for most elements in flames of about 3,000°K are in the 10^{-10} to 10^{-4} range. Thus, the atoms are predominantly in the unexcited or ground state. To get an absorption signal as trong as possible, AAS utilizes the resonance lines of those atoms in the ground (or near-ground) state. One can see that the amount of emission at any particular wavelength is affected considerably by changes in temperature, whereas atomic absorption is little affected since the number of atoms in the ground state remains constant. Consider the 422.6nm Ca line at 3,000 K, n^*/n is about 3.7 X 10^{-5}, whereas at 2,000°K, it is 1.2 X 10^{-7}. There are about;, 100 times more atoms in the excited state at higher temperature and hence capable of emission, whereas the perce0ntage of atoms in the ground state goes from 99.999 to 99.99999%, representing essentially no change at all. Nonetheless, the temperature of the flame does play an important role in AAS, since the amount of ionization and the number of molecules (e.g., $A1_2O_3$ rather than Al) present are directly dependent on it. Both ionization and molecular formation remove atoms from the flame and thus decrease the amount of absorption. For a more complete treatment of the theory basic principles and history, see Dean and Rains (1969), Winefordner (1976), Price (1979) and Kirkbright and Sargent (1974).

Appendices

Appendix-I: Terminology

Standard Solution: The solution of accurately known strength (or concentration) is called standard solution. It contains a definite number of gram equivalent or gram moles per litre of solution. Strength of a solution refers to the weight of a solute dissolved in a unit weight of the solution. It can be expressed in many ways as follows –

Normal Solution and Normality : A normal solution is one which contain 1 gram equivalent (eq.wt. in grams) of the active reagent, dissolved in 1 litre of the solution. Normality is the number of gram equivalent of the substance, dissolved in 1 litre of the solution. If the number of gram equivalent is 1, it is expressed as 1 N. If the number is 1/10, 1/100 or 1/1000 then it is designated as 0.1 N (decinormal), 0.01 N (centinormal) or 0.001 N (millinormal) solution respectively.

Normality = (No. of gram eq. of solute) / (No. of litre of solution)

Molar Solution and Molarity : A molar solution is one which contain a gram of molecular weight of the solute dissolved in 1 litre of solution. It is denoted by M. Whereas, molarity is the number of gram molecules of the substance dissolved in 1 litre of the solution.

Molal Solution : A molal solution is the one which contain a number of gram molecules of the solute dissolved in 1000 g of water.

Percentage composition by weight : The concentration is expressed in terms of the gram of solute per 100 g of solution. e.g. 10% KCl solution is prepared by dissolved 10 g of the salt in 90 g of water.

Percentage composition by volume : The concentration is expressed in terms of volume of the solute and solvent. e.g. 25 g of solution of methanol is prepared by mixing 25 ml of methanol with 80 ml of water.

Parts per million (ppm) : The concentration is expressed in terms of grams of solute per million millilitres of solution or milligrams of solute per litre of the solution. Thus a solution containing 10 mg / litre of solute or 10 microgram of solute per millilitre or solution is 10 ppm solution.

Milli equivalent per litre : A solution containing milli (1/1000) g equivalent of substance in a litre of the solution is expressed as meq/litre.

Titration : titration is defined as the process of determining the volume of a substance required to just complete the reaction with a known amount of other substance. The quantitative analysis carried out by titration is known as titrimetric analysis.

a) **Titrant:** The solution of accurately known strength used in titration is called titrant.

b) **Titrate:** The substance (in solution) to be determined by titration is called titrate.

Appendix-II: Molecular and Equivalent Weights of Some Important Compounds

Compound	Formula	Mol. Wt. (g)	Eq. wt. (g0
Ammonium acetate	CH_3COONH_4	77.08	77.08
Ammonium chloride	NH_4Cl	53.49	53.49
Ammonium fluoride	NH_4F	37.04	37.04
Ammonium nitrate	NH_4NO_3	80.04	80.04
Barium acetate	$(CH_3COO)_2$ Ba	255.43	127.72
Barium chloride	$BaCl_2.2H_2O$	244.28	122.14
Boric acid	H_3BO_3	61.83	20.61
Calcium acetate	$(CH_3COO)_2$ Ca	158.00	79.00
Calcium carbonate	$CaCO_3$	100.09	50.05
Calcium chloride (dehydrate)	$CaCl_2.2H_2O$	147.02	73.51
Calcium hydroxide	$Ca(OH)_2$	74.00	37.00
Calcium nitrate	$Ca(NO_3)_2$	164.00	82.00
Calcium sulphate	$CaSO_4.7H_2O$	172.17	86.08
Ferrous ammonium sulphate	$(NH_4)_2SO_4.FeSO_4.6H_2O$	392.13	196.06
Ferrous sulphate	$FeSO_4.7H_2O$	278.01	139.00
Magnesium chloride	$MgCl_2.6H_2O$	203.30	101.65
Magnesium nitrate	$Mg(NO_3)_2.\ 6H_2O$	256.41	128.20
Oxalic acid	$C_2H_2O_42H_2O$	126.00	63.00
Potassium chloride	KCL	74.55	74.55
Potassium dichromate	$K_2Cr_2O_7$	294.19	49.04
Potassium hydroxide	KOH	56.10	56.10
Potassium permagante	$KMnO_4$	158.03	31.60
Potassium nitrate	KNO_3	101.10	101.10
Potassium sulphate	K_2SO_4	174.27	87.13
Potassium hydrogen phthalate	COOH C_6H_4 COOK	204.22	204.22
Silver nitrate	$AgNO_3$	169.87	169.87
Sodium acetate (anhydrous)	CH_3COONa	82.04	82.04

Compound	Formula	Mol. Wt. (g)	Eq. wt. (g0
Sodium bicarbonate	$NaHCO_3$	84.01	84.01
Sodium carbonate	Na_2CO_3	106.00	53.00
Sodium chloride	NaCl	58.45	58.45
Sodium hydroxide	NaOH	40.00	40.00
Sodium nitrate	$NaNO_3$	84.99	84.99
Sodium oxalate	$Na_2C_2O_4$	134.00	67.00
Sodium sulphate	Na_2SO_4	142.04	71.02
Sodium thiosulphate	$Na_2S_2O_3.5H_2O$	248.18	248.18

Appendix-III: Conversion Factors for SI and Non-SI Units

To Convert Column 1 into column 2, multiply by:	Column 1 SI Unit	Column 2 non-SI Unit	To convert Column 2 into Column 1, Multiply by:
Length			
0.621	Kilometre.km (10^3m)	Mile,mi	1.609
1.094	meter, m	Yard, yd	0.914
3.28	meter, m	Foot, ft	0.304
1.0	Micrometer, µm (10^{-6} m)	Micron	1.0
3.94 x 10^{-2}	Milimeter, mm (10^{-3} m)	Inch, in	25.4
10	Nanometer, nm (10^{-9} m)	Angstrom, Å	0.1
Area			
2.47	Hectare, ha	Acre	0.405
247	Square kilometre, km^2 (10^3 m)^2	Acre	4.05 x 10^{-3}
0.386	Square kilometre, km^2 (10^3 m)^2	square mile, mi^2	2.590
2.47 x 10^{-4}	Square meter, m^2	Acre	4.05 x 10^3
10.76	Square meter, m^2	Square foot, ft^2	2.29 x 10^{-2}
1.55 x 10^{-3}	Square millimetre, mm^2 ($10^{-3}m^2$)	Square inch, in^2	645
Volume			
9.73 x 10^{-3}	Cubic meter, m^3	Acre-inch	102.8
35.3	Cubic meter, m^3	Cubic foot, ft^3	2.83 x 10^{-2}
6.10 x 10^4	Cubic meter, m^3	Cubic inch, in^3	1.64 x 10^{-5}
3.53 x 10^{-2}	Liter, L (10susp- m$^{3)}$	Cubic foot, ft^3	28.3
0.265	Liter, L (10^{-3} m^3)	Gallon	3.78
33.78	Liter, L (10^{-3} m^3)	Ounce (fluid), oz	2.96 x 10^{-2}
2.11	Liter, L (10^{-3} m^3)	Pint (fluid, pt	0.473
Mass			

To Convert Column 1 into column 2, multiply by:	Column 1 SI Unit	Column 2 non-SI Unit	To convert Column 2 into Column 1, Multiply by:
2.20×10^{3}	Gram, g (10^{-3}g)	Pound, lb	454
3.52×10^{-2}	Gram, g (10^{-3} kg)	Ounce (avdp), oz	28.4
2.205	Kilogram, kg	Pound lb	0.454
0.01	Kilogram, kg	Quintal (mtric), q	100
1.10×10^{-3}	Kilogram, kg	Ton (200lb), ton	907
1.102	Megagram, mg (tonne)	Ton (U.S.)ton	0.907
1.102	Tonne, t	Ton (U.S.) ton	0.907
Yield and Rate			
0.893	Kilogram per hectare, kg ha^{-1}	Pound per acre, lb $acre^{-1}$	1.12
0.107	Liter per hectare, L ha^{-1}	Pound per acre	9.35
893	Tonnes per hectare, t ha^{-1}	Pound per acre, lb $acre^{-1}$	1.12×10^{-3}
893	Megagram per hectare, Mg ha^{-1}	Pound per acre, lb $acre^{-1}$	1.12×10^{-3}
0.446	Megagram per hectare, Mg ha^{-1}	Ton (2000lb) per acre, ton $acre^{-1}$	2.24
Pressure			
9.90	Megapascal, MPa (10^{6} Pa)	Atmospher	0.101
10	Megapascal, MPa (10^{6} Pa)	Bar	0.1
1.00	Megapascal, per cubic meter, Mg M^{-3}	Gram per cubic centimetre, g cm^{-3}	1.00
2.09×10^{-2}	Pascal, Pa	Gram per square foot, lb ft^{-2}	47.9
1.45×10^{-4}	Pascal, Pa	Gram per square inch, lb in^{-2}	6.90×10^{3}
Temperature			
1.00 (K-273)	Kelvin, K	Celsius, ^{0}C	1.00 (^{0}C + 273)
(9/5 ^{0}C) + 32	Celsius, ^{0}C	Fahrenheit, ^{0}F	5/9 (^{0}F-32)
Electrical conductivity, electricity, and Magnetims			
10	Siemen per meter, S m^{-1}	Millimho per centimetre, mmho cm^{-1}	0.1
10^{4}	Tesl, T	Gauss, G	10^{-4}
Water Measurement			

To Convert Column 1 into column 2, multiply by:	Column 1 SI Unit	Column 2 non-SI Unit	To convert Column 2 into Column 1, Multiply by:
9.73 x 10^{-3}	Cubic meter, m^3	Acre-inches, acre-in	102.8
9.81x 10^{-3}	Cubic meter per hour, m^3h^{-1}	Cubic feet per second, $ft^3\ s^{-1}$	101.9
4.40	Cubic meter per hour, m^3h^{-1}	US. Gallons per minuter, gal min^{-1}	0.227
8.11	Hectare-meter, ha-m	Acre-feet, acre-ft	0.123
97.28	Hectare-meters, ha-m	Acre-inches, acre-in	1.03 x10^{-2}
8.1 x 10^{-2}	Hectare-centimeters, ha-cm	Acre-feet, acre-ft	12.33
Concentrations			
1	Centimole per kilogram, cmol kg^{-1}	Milliequivalents per 100 grams.	1
	(ion exchange capacity)	Meq $100g^{-1}$	
0.1	Gram per kilogram, g kg^{-1}	Percent, %	10
1	Milligram per kilogram, mg kg^{-1}	Parts per million, ppm	1

Appendix-IV: Conversion Factors for Different Elements

P_2O_5 x 0.44 = P	MgO x 0.60 = Mg	$CaCO_3$ x 0.40 = Ca
K_2O x 0.83 = K	S x 3.00 = SO4	$CaCO_4$ x 0.29 = Ca
P x 2.29 = P_2O_5	SO_4 x 0.33 = S	$MgCO_3$ x 0.28 = Mg
K x 1.20 = K_2O	HNO_3 x 0.22 = N	$MgSO_4$ x 0.20 = Mg
N x 1.22 = NH_3	H_3PO_4 x 0.32 = P	H_2SO_4 x 0.33 = S
Ca x 1.40 = CaO	$Ca_3(PO_4)$ x 0.20 = P	$CaSO_4$ x 0.24 = S
CaO x 0.71 = Ca	KCl x 0.52 = K	Na x 2.305 = Na2CO3
Mg x 1.67 = MgO	K_2SO_4 x 0.45 = K	

Appendix-V: Generalized Guidelines for Interpretation of Soil Analysis Data

Measurement	Soil Test Method	Low	Marginal	Adequate
Organic matter	Walkley-Black	<0.86 %	0.86-1.29%	>1.29
Organic Carbon		<0.5	0.50-0.75	> 0.75
Nitrate phosphate	AB-DTPA	<11	11-20	>20
	$NaHCO_3$	<8	8-15	>15

Measurement	Soil Test Method	Low	Marginal	Adequate
Potassium	AB-DTPA	<4	4-7	>7
	NH^4OA_C	<100	100-150	>150
Zinc	AB-DTPA	<60	60-120	>120
	DTPA	<0.5	0.5-1.0	>1.0
Copper	AB-DTPA	<1.0	1.0-1.5	>1.5
	DTPA	<0.2	0.2-0.5	>0.5
Iron	AB-DTPA	<0.2		>0.5
	DTPA	<4.5		>4.5
Manganese	AB-DTPA	<2.0	1.1-4.0	>4.0
	DTPA	<1.0	1.0-2.0	>2.0
Boron	AB-DTPA	<1.8		>1.8
	Hot water	<0.5	0.5-1.0	>1.0
	HCl	<0.45	0.45-1.0	>1.0

DTPA- diethylene triamine pentacetic acid. AB= ammonium bicarbonate.

$NaHCO_3$– Sodium bicarbonate.

Appendix-VI:Rating* Chart for Soil Test Value of Primary Nutrients

Nutrient	Rating*		
	Low	Medium	High
Organic Carbon (%)**	<0.5	0.50-0.75	>0.75
Alkaline $KMnO_4$ – N (kg ha^{-1})	<280	281-560	>560
Olsen's P (kg ha^{-1})	<10	11-25	>25
Amm. Acetate – K (kg ha^{-1})	<120	121-280	>280
S	<10	10-15	>15
Zn	<0.6	0.6-1.2	>1.2
Fe	<4	4-8	>8
Cu	<0.2	0.2-0.4	>0.4
Mn	<2	2-4	>4
B	<0.25	0.25-0.5	>0.5

Appendix-VII:Procedures for Preparing 1000 mg L^{-1} Standard Solutions*

Element	Procedure for making 1 L of Standard solution
Aluminium	Dissolve 1.0000g of Al metal in 25 mL of conc. HCl and a few drops of conc. HNO_3

Calcium	Dissolve 2.4973g of $CaCO_3$ (dried at 150^0C) in 25 mL of 1*N* HCl
Cadmium	Dissolve 1.0000 g of Cd metal (99.99%) in 50 mL of 1:1 HCl
Cobalt	Dissolve 1.0000 g of Co metal (99.99%) in minimum of 6 *N* HNO_3 or dissolve 2.630 g of $CoSO_4$ (dried at $250\text{-}300^0C$ for 6-8 hrs.) in about 100 mL of double distilled water and 1 mL of conc. H_2SO_4 (AR grade)
Chromium	Dissolve 1.0000 g of Cr metal (99.99%) in 50 mL of conc. HCl or Dissolve 2.8282 g of $K_2Cr_2O_7$ in double distilled water
Copper	Dissolve 1.0000 g of Cu metal (99.99%) in 50 mL of 1:1 diluted HNO_3
Iron	Dissolve 1.0000 g of Fe metal (99.99%) in 100 mL of 3.5 N H_2SO_4
Potassium	Dissolve 1.9067 g of dried KCl in double distilled water
Magnesium	Dissolve 1.0000 g of Mg metal ribbon or turning (99.99%) in 50 mL of 5*N* HCl.
Manganese	Dissolve 1.0000g of Mn metal wire or foil (99.99%) in 50mL of 6*N* HNO_3 or dissolve 3.076 g of manganous sulphate monohydrate ($MnSO_4.H_2O$) in 200 mL of double distilled water and add 1.5 mL of conc. HNO_3 or dissolve 1.5824 g of manganese dioxide (MnO_2) in minimum quantity of conc. HCl, evaporate to dryness, dissolve residue in double distilled water
Molybdenum	Dissolve 1.5003 g of molybdenum trioxide (MoO_3) in 10 mL of conc. HCl
Nickel	Dissolve 1.0000 g of Ni metal (99.99%) in 50 mL of 1:1 diluted HNO_3
Lead	Dissolve 1.0000 g of Pb metal (99.99%) in 20 mL of 6*N* HNO_3 or Dissolve 1.5982 g of Pb $(NO_3)_2$ in DDW and add 20 mL of conc. HNO_3
Selenium	Dissolve 1.0000 g of Se metal (99.99%) in 20 mL of aqua-regia (15 mL conc. HCL + mL conc. HNO_3)
Silicon	Fuse 4.278 g of silicon dioxide (SiO_2) with 20 g of sodium carbonate in a platinum crucible and dissolve in distilled water.
Zinc	Dissolve 1.0000 g of Zn metal (99.99%) in 50 mL of 1:1 diluted HCl or dissolve 4.5490 g of zinc nitrate hexahydrate [Zn $(NO_3)_2$ $6H_2O$] in double distilled water.

Appendix-VIII: Preparation of 1000 ppm Solutions of Different Elements

Sr. No.	Elements	Elements Quantity of salt to be dissolved in 1000 ml distilled water
1	Nitrogen	4.719 g of $(NH_4)_2SO_4$ Ammonium sulphate
		2.140 g urea $CO(NH_2)$ 2
2	Phosphorus	4.390 g of KH2PO4 Pot. Dihydrogen phosphate
		5.555 g of CaHPO4.2H2O, Dicalcium phosphate dihydrate
		14.31 g of single superphosphate containing 6.986% P
3	Potassium	1.906 g Potassium chloride (Muriate of potash) KCl
4	Zinc (Zn)	4.398 g Zinc sulphate ZnSO4.7H2O
5	Manganese (Mn)	3.609 g Manganese chloride, MnCl2.2H2O
		3.077 g Manganese sulphate MnSO4.H2O
6	Copper (Cu)	3.929 g Copper sulphate CuSO4.5H2O

7	Iron (Fe)	4.978 g $FeSO_4.7H_2O$, Ferrous sulphate
		7.021 g $(NH_4)_2.Fe(SO_4)_2.6H_2O$ of Ferrous ammonium sulphate
8	Molybdenum	12.139 g Ammo. Molybdate $(NH_4)_6Mo_7O_{24}.4H_2O$
		2.521 g of Sodium molybdate $Na_2MoO_4.2H_2O$
9	Boron	5.720 g Boric acid H_3BO_3

Appendix – IX: Some Indicator Plants of Nutrient Deficiency

Nutrients	Indicator Plants
N	Maize, cereal (small grain), mustard, apple, citrus
P	Maize, barley, lettuce, tomato.
K	Potato, lucern, beans, tobacco, cucurbits, cotton,
Ca	tomato, maize.
Mg	Lucern, other legume crops.
S	Potato, cauliflower.
Fe	Lucern, raya.
Zn	Sorghum, barley, citrus, peach.
Mn	Maize, onion, citrus, peach.
Cu	Apple, cherry, citrus, maize, oats, pea, radish, wheat.
B	Apple, citrus, barley, maize, lettuce, oats, onion,
Cl	tobacco, tomato,
Mo	Lucern, turnip, cauliflower, apple, peach.

References & Suggested Readings

Abrol, I.P.; Yadav, J.S.P and Masoud, Fl. 1988. Salt Affected Soils and Their Management. FAO Soils Bull. 39.

Baethgan, WE. and Alley, M.M. 1989. A manual colorimetric method of measuring ammomum N in soil and Kjeldabl digest. Comm. Soil Sci. Plant Anal. 20: 96 1-969.

Bray, R.H. and Kurtz, Li 1945. Determination of total, organic and available forms of phosphorus in soils. Soil Sci. 59: 39-45.

Datta, N.P.; Khera, M.S. and Saini, TR. 1962. A rapid colorimetric procedure for the determination of the organic carbon in soils. J. Indian Soc. Soil Sci. 10: 67-74.

Dean, J.A. and Rains, IC. 1969. Flame emission and atomic absorption spectrophtometry, Vol 1. Theory, Marcel Dekker, Inc., New York.

FAO. 1980. Soil testing and plant analysis.Bull. No. 38/1, Food and Agriculture Organisation, Rome, Italy.

Gupta. PK. 1999. Soil, Plant, ter and Fertilizer Analysis. AGROBIOS (India) Reprint (2000). 438 p.

Hanway, J.J. and Heidel, ff1952. Soil analysis methods as used in Iowa State College Soil Testing Laboratory. LowaAgric. 57: 1-3 1.

Hesse, ER. 1994. A Text Book of Soil Chemical Analysis. CBS Publishers, Delhi.

Jackson, M.L 1958. Soil Chemical Analysis. Prentice Hall, New Jersey, USA.

John, M.K.; Chuah, H.H. and Ndufeld, J.H. 1975. Application of improved azomethine- H method to the determination of boron in soils and plants. Anal. Lett. 8: 559-68.

Johnson, C.M. and Arkley, TH. 1954. Determination of molybdenum in plant tissue. Anal. Chem. 26: 572-574.

Kadam, J. R., Shinde P. B., 2005, Practical Manual on Soil Physics – A method manual, Department of Agricultural Chemistry and Soil Science, P.G.I., Rahuri, P-24.

Katyal, J.C. and Rattan, RJC 2003. Secondary and micronutrients- Research gaps and future needs. Fert. News 48(4):14-20.

Kirkbright, G.E and Sargent, M. 1974. Atomic absorption and fluorescence spectroscopy. Academic Press, Inc., New York.

Knudsen, D., Paterson, G. A., and Pratt, P2.1982. Lithium sodium and potassium. Pages 225-246 inA.L Pageetal,Eds. Methods of soil analysis. Part 2; 2nd ed. Agronomy No.9 American Society of Agronomy, Madison, WI. -. Under,

lkley, AJ. and Black, LA. 1934. Estimation of soil organic carbon by the chromic acid titration method. Soil Scil. 37: 29-38.

Mahapatra, B.S. 1990. A practical Manual for Soil Fertility Management. Deptt. of Agronomy, College of Agriculture, G.B. Pant Univ. of Agric. & Tech., Pantnagar, Uttaranchal, India.

Olsen, S.R., Cole, C.V., Watanabe, ES. and Dean, L.A. 1954. Estimation of available phosphorus in soils by extraction with sodium bicarbonate. Cfrc. U.S. Dep. Agric. 939.

Prasad, R. 1965. Determination of potentially available nitrogen in soils - A rapid procedure. PL Soil 23:261-263.

Prasad, R., Shivey, YS., Kitmar, D. and Sharma, S.N. 2006. Learning by Doing Exercises in Soil Fertility (A practical Manmil for Soil Fertility), Division of Agronomy, IARJ, New Delhi, 68p.

Price, WZ.1979. Spectrochemical analysis by atomic absorption. Hey&n & Son, Ltd., London. Purvis, E.R. and Peterson, N.K. 1965. Methods of soil and plant analysis for molybdenum. Soil Sci. 81:223-228.

R.C. and Harley, C.P. 1942. A rapid method for the determination of nitrogen in plant tissue. Science 96: 565-566.

Schneider Karl R. 1996. Soil Testing and Plant Analysis for Fertilizer Recommendations, Quick Bibliography series no. Q B 96-12 (March 1993- June 1996), 148 citation from ARICOLA Database sep. 1996.

Schoonover, WR. 1952. Examination of soils for alkali. Bull. Univ. California Ext. Ser., Berkeley (mimeographed publication ci. Singh et al., 1999).

Singh, D., Chhonkar, EL and Pandey, RN. 1999. Soil, Plant, Water Analysis-A Methods' Manual, Division of Soil Science and Agricultural Chemistry, IARI, New Delhi.

Soil and Plant Analysis Council. 1992. Hand Book on reference methods for soil analysis, Aihense, GA, USA.

Soil Testing Procedure Manual 2008, Marathwada Agricultural University, Parbhani, P-30.

Subba Rao, A. 1995. Analysis of soils for available major nutrients. In: H.LS. Tandon (ecL).' Methods of Analysis of Soils, Plants, Waters and Fertilizers. FDCO, New Delhi, p. 13-35.

Subbiab, B.V. and Asija, G.L 1956. A rapid procedure for assessment of available nitrogen in rice soils. CUff. SciL 25: 259-260.

Tandon, H.LS. 1987. Phosphorus Research and Agriculture Production in India. Fertilizer Development and Consultation Organisation, New Delhi.

Tandon, H.LS. 1993. Method of Analysis of soil, Plants, Waters and Fertilizers, (ed) Fertilizer Development and Consultation Organization, New Delhi.

Tekalign, T Hague I and Aduayi E.A.1991. Soil, Plant, Water, fertilizer, animal manure and comport analysis manual. Plant Science Division Working Document 13. ILCA, Addis Ababa, Ethiopia.

Williams, C.H. and Steinbergs, A. 1959. Soil sulphur fractions as chemical indices of available sulphur in some Australian soils. Aust. J. agric. Res. 10: 340-52.

Wmefordner, J.D. (ed.). 1976. Trace analysis, spectroscopic methods for elements, Vol. 46. Chemical analysis (a series of monographs on analytical chemistry and its applications). John Wiley & Sons, Inc., New York.